**Economic
Disincentives
for Energy
Conservation**

Economic Disincentives for Energy Conservation

Joe W. Russell, Jr.

Environmental Law Institute State and Local Energy
Conservation Project

This work was supported in part by a grant from the National Science Foundation.

Ballinger Publishing Company • Cambridge, Massachusetts
A Subsidiary of Harper & Row, Publishers, Inc.

This book was prepared with the support of NSF Grant APR 7504814. However, any opinions, findings, conclusions, or recommendations herein are those of the authors and do not necessarily reflect the views of NSF.

International Standard Book Number: 0-88410-060-X

Library of Congress Catalog Card Number: 79-13170

Printed in the United States of America

Library of Congress Cataloging in Publication Data

Russell, Joe W., Jr.
  Economic disincentives for energy conservation.

  Includes bibliographical references and index.
  1. Energy policy—United States. 2. Energy conservation—
Law and legislation—United States. 3. United States—Economic
policy—1971- I. Environmental Law Institute. Energy Conservation Project. II. Title.
HD9502.U52R87   333.7   79-13170
ISBN 0-88410-060-X

Dedication

To: Kathy

Contents

List of Tables

Preface

At one time, Americans believed that there was an energy crisis: long gas lines and closed factories demonstrated that something had gone wrong. The automatic process of providing cheap energy when and where we wanted it had broken down. Yet memories of the crisis faded quickly, to be replaced with suspicion that perhaps the emergency had been manufactured by government incompetence or private greed. The sense of urgency had so far departed that within three years after the embargo, a majority of Americans didn't believe that there was an energy crisis, didn't know that we imported oil, and thought that a technological fix would solve the supply shortages.

In the face of this strongly held belief that everything would work out, there was one anomaly: home insulation sales soared to record heights. Storm windows, attic batts, water heater insulating jackets—all were added to the homeowner's shopping list. What explains this split between perception and action? When citizens no longer believed their government's words about the energy situation, nonetheless they acted. What had caught their attention? The answer is obvious. The price of energy had risen, so that it became a good investment to conserve. Insulation sales were a dramatic illustration of the fact that when price talks, people listen.

Since 1975, the Environmental Law Institute has been studying the opportunities that state and local governments have for encouraging energy conservation. The Institute's Energy Conservation Project, funded under grants from the National Science Foundation, has produced a series of books designed to assist citizen leaders and state and local legislators and government officials in selecting and implementing promising conservation strategies. This book examines a novel—and almost certain to be controversial—type of conservation strategy, the economic disincentive. Economic disincentives (taxes, charges, or fees imposed by the government) take advantage of the fact that people will pay attention when the price of energy rises. Imposed on energy itself or on energy-consuming goods and activities, economic disincentives would be a powerful force for redirecting this nation from a fossil-fuel-based economy to one based on the use of renewable resources.

Unlike other approaches to raising the price of energy, such as decontrolling energy prices, economic disincentives have the advantage that they would generate revenues for states and localities, rather than for the energy-supplying companies.

Use of economic tools to implement social policy is a relatively new phenomenon in this country. Aside from a few pollution control rules in Connecticut and some experimentation by the federal Environmental Protection Agency, Americans have much to learn about crafting economic disincentives to the tasks customarily borne by regulation. Jay Russell's book is an attempt to answer some of the difficult legal and practical questions in the context of saving energy. Looking at a variety of innovative strategies, he does not ignore the political difficulties of raising prices to low income consumers. Nor does the American sacred cow, the private automobile, escape his attention. This is clearly a bold book, one that will provoke discussion and debate in the state capitals. Whether any state or local legislator will rise to the challenge will be one significant test of our commitment to saving energy.

> **Grant P. Thompson**
> Institute Fellow
> Principal Investigator
> Energy Conservation Project

Acknowledgments

Many people have helped bring this book to fruition, but foremost among those who deserve credit is my wife, Kathryn L. Russell, Esq. As the book neared completion, she provided research assistance and suggested many substantive and stylistic changes to improve the manuscript. From the beginning, she sustained me with her patience, support, and love.

Special thanks go to Grant P. Thompson, who directed the Energy Conservation Project, for giving me the opportunity to write the book, and to Gail Boyer Hayes, who currently directs the Environmental Law Institute's energy program, for encouraging me with her faith in my ability to do it.

I would also like to thank my former colleague at the Institute, E. Grant Garrison, who co-authored an early draft of the book.

Many people reviewed drafts of various chapters and offered numerous helpful comments: Ohio State Representative Sherrod Brown, Richard Ebel of the District of Columbia Tax Revision Commission, and Lee Lane of the Public Interest Economics Center, all of whom served on the project's National Advisory Panel for this book; several Institute economists and attorneys, including Stephen O. Andersen, Robert Anderson, Will A. Irwin, and Philip D. Reed; Thomas G. Krattenmaker and William H. Rodgers, Jr., professors at Georgetown University Law Center; David W. Tundermann of the President's Council on Environmental Quality; Hazel R. Rollins, general counsel of the Community Services Administration; and Joel Sharkey of the Michigan Public Service Commission.

Cheryll Williams and Kathy Yaksick, who typed the final draft, and Margaret Hough, who edited the rough edges, have my gratitude for their assistance in preparing the manuscript for publication.

Finally, I want to thank my friends from the Energy Conservation Project for making our joint undertaking an experience I shall never forget.

Introduction

Today, the energy crisis is over. Taking its place is a less obvious but more intractable and insidious energy problem. Consequently, in the years ahead, the energy situation will continue to present the greatest domestic challenge to the nation's will and imagination since the Great Depression. How the nation responds (or fails to respond) to this challenge will affect for better (or worse) the economic well-being and lifestyle of every American.

From the start of the Arab oil embargo in 1973, the country has debated what approaches individuals, businesses, and government should adopt to deal with the energy problem. So far, most of the public's attention has centered on federal efforts to develop and implement a national energy policy. The Environmental Law Institute State and Local Energy Conservation series hopes to contribute to the continuing energy debate by focusing attention on the important role that state and local governments have in carrying out one of the key elements of national policy—encouraging energy conservation. The books in this series, therefore, examine a variety of energy conservation strategies that are within the powers of the states and localities to enact.

The need for state and local government initiatives to encourage conservation arises from the persistent failure of the public to take advantage of many reasonable conservation opportunities. This failure has occurred at least partly because the public has lacked sufficient economic motivation to save energy. In other words, the cost of energy (prices plus taxes) in this country has remained too low, even after prices jumped as a result of the embargo, to convince individuals to cut their energy use significantly.

Taking its inspiration from an observation by Justice Louis D. Brandeis that "[i]t is one of those happy incidents of the federal system that a single courageous State may, if its citizens choose, serve as a laboratory; and try novel social and economic experiments without risk to the rest of the country,"[1] this book analyzes one of the options available to states and localities for supplying a stronger economic motivation for energy conservation—the imposition of economic disincentives. Economic disincentives are monetary charges levied by the government for the principal purpose of discouraging energy-wasteful conduct.[2] As promising, but relatively novel and basically untried, measures, economic disincentives merit testing to determine their actual effectiveness.

The purpose of this book is to assist states and localities that choose to experiment with economic disincentives. This book does not, however, examine every conceivable economic disincentive. Some strategies that could be described as economic disincentives or that would have the same result (i.e., raising the cost of energy to encourage conservation) are not given detailed treatment here because they are discussed in other books in the series. Among these strategies are beverage container deposits ("bottle bills"), product charges, and electric utility rate reform.[3] Disincentives such as a general energy or BTU tax that seem to have no realistic chance for enactment at the state or local level have also not been included. In the case of a BTU tax, it seems unlikely that states would impose a tax to save energy if the tax has the potential for driving out existing, or keeping out new, business and industry. Like air and water pollution regulations, a BTU tax seems more appropriately a subject for national, rather than state, action. Several pollution charges, such as a tax on lead additives in gasoline, were studied as potential disincentives, but have not been included in this book because of their limited impact on energy consumption.[4] The possibility of achieving more energy-efficient land use patterns through economic disincentives was also investigated, but neither approach examined (differential assessment and taxation and land gains taxation) offered a clear indication of success.[5] Finally, other economic disincentives, such as appliance efficiency taxes and rebates, were considered but not pursued in instances where existing government energy programs appear to be successfully encouraging people to save energy.[6] Rather than devise strategies that may not be needed, this book chose to focus instead on disincentives for those problem areas in which the opportunities for conservation have not been fully realized.

Although this book does not provide a complete catalog of economic disincentives, it does provide a framework that can be used to analyze any disincentive proposal. The framework emphasizes the legal, policy, and to a lesser extent, economic considerations and implications that states and localities should weigh in deciding whether to impose these measures. Chapter 1 lays the foundation for the framework by discussing the nation's energy problem, the role of energy conservation in dealing with the problem, and the relationship of economic disincentives to energy conservation. Chapter 2 examines the major advantages and disadvantages of disincentives and the chief obstacles to their adoption. Chapter 3 explores the legal and policy issues that should be considered as specific disincentives are designed. The framework is applied in Chapter 4, which analyzes several disincentives aimed at breaking up the love affair between Americans and their cars without seriously impairing personal mobility. Chapter 5 concludes the book by comparing measures for alleviating the burden of increasing residential energy costs on the poor and the elderly.

NOTES TO INTRODUCTION

1. New State Ice Co. v. Liebmann, 285 U.S. 262, 311 (1932) (Brandeis, J., dissenting).

2. This definition is derived from the definition of economic disincentives in W.A. Irwin and R.A. Liroff, *Economic Disincentives for Pollution Control: Legal, Political and Administrative Dimensions*, EPA—600/5—74—026, prepared for the U.S. Environmental Protection Agency by the Environmental Law Institute (Washington, D.C.: U.S. Government Printing Office, 1974). A disincentive, according to Irwin and Liroff, is "a monetary charge levied by government on conduct which is not illegal but which does impose social costs, for the principal purpose of discouraging the conduct" (*id.* at 6).

3. Beverage container deposits ("bottle bills") and product charges are analyzed in N.L. Dean, *Energy Efficiency in Industry* (Cambridge, Massachusetts: Ballinger Publishing Co., 1979). Electric utility rate reform is treated in F.J. Wells, *Utility Pricing and Planning—An Economic Analysis* (Cambridge, Massachusetts: Ballinger Publishing Co., 1979).

4. For a detailed analysis of pollution charges, see F.R. Anderson et al., *Environmental Improvement through Economic Incentives* (Baltimore: The Johns Hopkins University Press, 1977).

5. The differential assessment and taxation approach is reflected in the Montana Economic Land Development Act, Mont. Rev. Codes Ann. §§ 84—7501 to 84—7526. In 1977, Montana abandoned efforts to implement this incredibly complex and confusing statute. The Vermont land gains tax, Vt. Stat. Ann. tit. 32, §§ 10001—10010, was examined to determine whether land value increment taxation in rural areas would, by detering land speculation, curb urban sprawl.

6. Federal residential energy conservation programs should, according to one analysis utilizing an engineering and economic model of residential energy use, achieve substantial energy savings—46 quads (1 quad = 10^{15} BTUs)—between now and 2000. E. Hirst and J. Carney, "Effects of Federal Residential Energy Conservation Programs," *Science* 199 (February 24, 1978): 845.

Chapter 1

The Energy Problem
and Energy Conservation

THE ENERGY SITUATION:
PAST, PRESENT, AND FUTURE

From the end of World War II until the fall of 1973, the United States luxuriated in an era of inexpensive and abundant energy supplies. Energy fueled strong economic growth and a rising standard of living, and America became hooked on cheap energy. When domestic sources of oil could no longer satisfy the nation's energy habit, imported oil filled the void. By 1972, the United States was importing nearly one-third of the oil it consumed. During this period, as America's appetite for energy grew enormously, few people worried whether energy was being wasted, whether nonrenewable resources were being depleted too rapidly, or whether an increasing dependence on foreign energy suppliers was wise. No one worried because there seemed no cause for concern.

In October 1973, however, the Arab oil embargo shocked the country out of its complacency. Not only did the Arab oil producers drastically cut their exports, but all oil-exporting nations raised prices. As a result, the cost of maintaining America's energy habit skyrocketed practically overnight. The term "energy crisis" was coined and came to signify all of the embargo's consequences. The public thought of the energy crisis primarily in terms of supply shortages, which citizens experienced as they sat in long lines at gasoline stations; higher prices for all types of energy; and the oil companies' record ("obscene") profits.

At first, the public responded to the energy crisis by searching for villains and conspiracies. Gradually, however, people began reducing their energy consumption. By the end of 1974, overall energy use had fallen 2.2 percent from the 1973 level, and the demand for oil had dropped 3.7 percent.[1] Government reacted to the crisis by attempting to alleviate the hardships caused by the sudden shortage of oil. Congress passed the Emergency Petroleum Allocation Act of 1973[2] to establish authority for allocating petroleum supplies and controlling oil prices. The states administered gasoline allocation plans, allowing motorists to buy fuel only on even- or odd-numbered days according to the last digits of their license plates.

After the Arabs lifted their embargo in the spring of 1974, the lines disappeared, prices slowed their rapid escalation, and consumers complained about, but accepted, higher prices as a fact of life. The supply emergency had passed, but the economic aftershocks of the crisis continued to rock the country. Foreign oil prices had quadrupled during the embargo, and by 1975 these higher prices had triggered the nation's worst recession since the Great Depression. As the nation prepared to enter a new energy era, the lesson of the embargo, which the recession served to reinforce, seemed clear: The United States had to control its energy habit because it could not afford the price exacted by its heavy dependence on foreign oil producers. Unfortunately, not everyone took this lesson to heart.

Government, however, recognized that the energy problem would not disappear if ignored, and Congress and the state legislatures began to consider and enact a variety of energy policies and programs.[3] Congress passed the Energy Policy and Conservation Act (EPCA) in late 1975 and the Energy Conservation and Production Act (ECPA)[4] eight months later. Neither law amounted to a comprehensive national energy policy, but both contained programs that were steps in the right direction.

Congress addressed several of the most pressing aspects of the energy problem in EPCA. To alleviate the effects of the high prices for foreign oil, Congress clamped price controls on domestic crude oil. To cushion the impact of a future embargo, the act authorized the president to prepare emergency energy conservation and rationing plans and established a Strategic Petroleum Reserve. To reduce the nation's reliance on foreign oil in the future, Congress set up several long term energy conservation programs. In addition to a voluntary program for the industrial sector, EPCA mandated improved automobile fuel economy and the testing and energy efficiency labeling of major home appliances. Most important for states and localities, though, were provisions authorizing federal assistance to states for the preparation and implementation of state energy conservation plans.[5]

Congress soon realized that EPCA's energy conservation programs failed to tap several areas of great energy conservation potential. In ECPA, Congress attempted to remedy one of EPCA's major omissions by establishing several programs to increase the energy efficiency of buildings. The programs ranged from one requiring the development of federal energy conservation performance standards for new residential and commercial buildings to several providing financial assistance to encourage energy conservation improvements in existing buildings. In another area of untapped energy conservation potential—electric utility rate reform—Congress only ordered a study.

While Congress prepared for the new energy era by passing EPCA and ECPA, the nation regained its voracious energy appetite. In 1975, the United States relied on foreign oil producers for 37 percent of the oil it consumed. In 1976, the nation imported 42 percent of its oil, and by the end of 1977, it was depending on foreign oil to satisfy nearly 50 percent of its needs. The growth in imports occurred, at least in part, because the economy was recovering from the severe recession that had restrained energy demand. The economic upswing, however, does not fully explain why, since 1976, gasoline consumption has soared to record levels every year. Several other factors account for the public's acting as if the energy crisis had never happened and as if no serious energy problem has existed since the embargo was lifted.

Many people have slipped back into their old energy habits because they do not know or understand that there has been and is an energy problem that affects them personally. Recent public opinion polls have exposed an astonishing degree of ignorance about America's energy situation. In the summer of 1977, three years after the embargo ended and only a few months after President Carter declared the "moral equivalent of war" on the energy problem, a CBS/*New York Times* poll discovered that 46 percent of the public did not know that the United States imports oil![6] A Gallup poll in the spring of 1978 revealed that the public had learned very little in the intervening months.[7]

The findings of these polls reflect not only the public's short memory and the failure of politicians to communicate and of the people to listen, but also the contrast between the 1973–1974 energy crisis and the subsequent energy problem. Unlike the conspicuous crisis precipitated by the embargo, the energy problem since 1976 has been largely invisible. The nation's massive dependence on foreign oil has not generated any dramatic symbols (e.g., lines at service stations) nor instilled a sense of existing or imminent emergency. News reports concerning temporary gluts of oil and natural gas in certain parts of the country have only confirmed the public's perception that there is no energy problem. Without any tangible evidence to the contrary, the country has ignored political rhetoric and made little effort to curb its energy habit.

Regardless of the public's perception, an energy problem does exist, and its ramifications touch virtually every American. Although oil from the North Slope of Alaska reduced imports slightly in 1978 (to 43 percent of demand during the first eleven months of the year), America remains too dependent on foreign oil. This reliance on imports leaves the nation highly vulnerable to another embargo that would undoubtedly be much more economically debilitating than the first.

Even if another embargo is unlikely, the nation should reduce its dependence on imported oil because of the highly detrimental effect that the current level of imports has on the economy. In 1973 the United States spent $3 billion on oil imports. By 1977 the bill had jumped to $42.4 billion. According to one estimate, it would take the proceeds from two years of American agricultural exports to recover the billions that were paid for foreign oil in 1977.[8] As a result of this dollar drain, the nation's trade deficit in 1977 reached a record level of $26.6 billion. The extent of the country's dependence on imported oil, the resulting trade imbalance, and the failure of the nation to enact an energy policy to reduce its imports all contributed to a loss of confidence in the soundness of the dollar. Beginning in late 1977, this loss of confidence was reflected in the ebbing value of the dollar in world money markets.

The public should not and cannot cavalierly ignore these complex matters of international finance in the belief that they have no significance for their day-to-day lives. As the value of the dollar abroad drops, American consumers not only pay more for foreign goods, but they eventually also pay more for everything else because a decline in the value of the dollar overseas aggravates inflation in the United States.[9] America's dependence on foreign oil also has an impact on domestic employment. Every $5 billion increase in oil imports costs this country 200,000 jobs.[10] Thus, exporting dollars to pay for the oil imported to satisfy the nation's energy habit adversely affects everyone's economic well-being.

Unfortunately, these grim statistics documenting the impact of the nation's appetite for foreign oil on the standard of living did not motivate either significant voluntary measures to break the energy habit or quick congressional action to enact a comprehensive national energy program. Although the energy crisis and the subsequent energy problem demonstrated the need for a national energy policy, congressional efforts embodied in EPCA and ECPA had fallen short. Recognizing the unfulfilled need, President Carter attempted to fill the void in April 1977 with his proposed National Energy Plan.[11] The objectives of the National Energy Plan included greater conservation of energy in general and of oil and natural gas in particular, increased production of domestic oil and gas, and a shift by industry and utilities from using oil and gas to coal. To accomplish these goals, the Plan proposed a variety of taxes, tax rebates, and tax credits; changes in the pricing policies for oil and natural gas; and regulatory measures.

After receiving the Plan from the President, Congress became bogged down in arguments over its controversial provisions for conservation taxes and natural gas pricing and delayed final action on all parts of the Plan until the arguments were resolved. Finally, in October 1978, Congress passed the National Energy Act (NEA),[12] a substantially different and weaker collection of measures than those the President originally proposed. The NEA consists of five parts: (1) energy tax provisions establishing tax credits for residential insulation and solar installation, business energy tax credits, and taxes on gas-guzzling automobiles; (2) a natural gas-pricing policy extending and continuing federal price regulations while allowing prices to rise gradually until 1985 when controls would expire for most categories of gas; (3) prohibitions on new oil- and gas-burning utility and industrial plants, requiring in effect that these plants use coal; (4) authorization for the federal government to intervene in state electricity rate proceedings to advocate various rate reforms that state utility commissions must "consider," but not necessarily adopt; and (5) a variety of conservation measures, including loans and grants to finance residential energy conservation improvements, mandatory energy efficiency standards for thirteen home appliances, and higher penalties for automobile manufacturers who fail to meet EPCA's fuel economy standards.

Lacking the conservation taxes proposed in the National Energy Plan, the NEA will not save as much energy as the Plan would have. The President predicted his Plan would save 4.5 million barrels of oil a day by 1985.[13] Estimates of the NEA's savings range from 2 million barrels (by congressional energy experts) to 2.4 to 3 million barrels (by the Department of Energy).[14]

Several factors explain why Congress rejected the tough energy program proposed by the President in favor of its softer version. Special interest groups lobbied heavily against the most effective measures in the Plan, and a rising national antitax fever made Congress reluctant to impose conservation taxes. Political caution also grew from a widespread perception that matters were perhaps not as grim as the President had painted them. Despite the massive trade deficit and the declining dollar, news stories about the glut of oil and natural gas made it difficult for legislators to convince themselves or their constituents that tough measures were necessary. Moreover, a rash of optimistic reports about the long term energy future undermined President Carter's assertion that the probability of a serious crisis by 1985 necessitated the enactment of the Plan.

Projections of future global and national energy supplies form the basis for long range energy planning, yet no consensus has developed as to what the future holds. Experts disagree, with the pessimists asserting that a shortage of oil and natural gas will occur in the 1980s and the optimists arguing that world reserves of oil and gas are sufficient to postpone the crunch until the 1990s or beyond. Recently, the optimists have gained the support of a number of respected authorities, including the Organization of Arab Petroleum Exporting Countries (OAPEC), the Trilateral Commission, and the Petroleum Industry Research Foundation. The optimists point to new estimates of world oil reserves, which indicate that reserves may be greater than previously thought, and to the production of oil and gas from reserves that have become economically feasible to exploit as a result of the increases in oil and gas prices over the last five years to buttress their contention that the world will experience neither a critical shortage of oil nor sharply higher prices until the mid-1990s.[15]

Unfortunately, the conflicting forecasts of the experts, even if time eventually proves the optimists correct, only confuse matters now. In a climate of uncertainty, the danger is that the nation will do nothing, postponing action perhaps until it is too late. The real question for decisionmakers is not which forecasts to believe, but whether the nation should prepare for the worst. By adopting energy policies now that are based on an assumption that the 1980s will bring another energy crisis, government will, in effect, be buying insurance that will pay off if the eventuality does occur. Even if the nation and the world do not experience severe shortages or price increases until the mid-1990s, conservation efforts will not have been wasted. They will have delayed the crunch even longer, and in any event, the nation will have had the benefits of being protected against potential disaster.

The crisis of the mid–1970s, the invisible problem of the late 1970s, and the uncertainty of the 1980s and 1990s will all shape the energy decisions that states and localities make in the next few years. Although state and local governments have several options to choose from in developing energy policies, policies emphasizing energy conservation will make a significant contribution toward achieving national energy objectives.

ENERGY CONSERVATION: WHAT, WHY, HOW MUCH, AND HOW

The federal government has a variety of strategies at its disposal for dealing with the many complex facets of the energy problem. State and local governments have fewer options from which to select. For example, states cannot relax price controls on oil and natural gas to encourage increased domestic production. One of the strategies, however, clearly within the power of the states and localities to pursue is energy conservation. Although the federal government has emphasized conservation policies, states and localities have numerous opportunities either to reinforce federal conservation programs or to break ground in areas that the federal government has neglected. In choosing either direction, state and local governments can have an almost immediate impact on the nation's energy problem.

What Is Energy Conservation?

Energy conservation should not be regarded as an end in itself, but rather as a means toward the end of improving the economic well-being of individuals, corporations, and the nation as a whole.[16] In other words, conservation does not involve saving BTUs for the sake of saving BTUs, but rather it involves saving BTUs when the economic rewards of saving them outweigh the costs incurred in saving them. Considering conservation from this perspective underlines the fact that saving energy means saving money.

Energy conservation can be achieved through both technological innovation and attitudinal and behavioral changes. The technological aspect of conservation focuses on finding technical answers to the problem of reducing the energy requirements of energy-consuming goods and processes. For example, government now requires automobile and appliance manufacturers to redesign their products to make them more energy-efficient. Industrial, manufacturing, and agricultural processes can be changed so that less energy is needed to produce output.[17] New residential and commercial buildings can be constructed to take advantage of a variety of energy-saving features, and existing buildings can be retrofitted to reduce their energy requirements.[18] Although it is crucial that the nation begin promptly on this technological shift, the results will not occur overnight.

In contrast, attitudinal and behavioral changes arising from a conservation ethic can produce more immediate energy savings. For example, people can save energy every day by being more careful in using household appliances, by walking or taking a bus for some trips instead of driving everywhere, by extinguishing unnecessary lights, and by turning the thermostat down in winter and up in the summer. In addition to these types of daily opportunities to reduce energy consumption, people have many other chances to practice energy conservation. For instance, people can conserve significant amounts of energy over time simply by purchasing the more energy-efficient models of various energy-consuming products. Developing a conservation consciousness and adjusting to energy-saving attitudinal and behavioral changes may initially impose some minor inconvenience, but this is a small price to pay for the substantial energy savings and economic benefits that will accrue from these efforts.

Finally, conservation must be distinguished from curtailment, a much more drastic strategy. Whereas conservation means that people drive more efficient cars and that factories utilize more energy-efficient processes, curtailment means severe restrictions or prohibitions on driving and closed factories when fuel supplies run short. Imposed during energy emergencies, curtailment measures would demand sacrifice and inflict economic hardship. In contrast, conservation does not involve self-denial or suffering and may help to forestall the necessity for resorting to curtailment strategies.

Why Conserve Energy?

The direct economic gains resulting from conservation efforts will serve as the primary motivation for individuals and corporations to save energy. Conservation, however, offers other benefits and advantages that provide additional incentives.

Conservation is the cornerstone of the nation's effort to reduce its dependence on imported oil. An earlier portion of this chapter details the deleterious economic effects of this reliance on foreign oil—fewer jobs, a weaker dollar, and higher inflation. By reducing oil imports, conservation should maintain or even raise the nation's standard of living.

Conservation is the least polluting, safest, and cheapest source of energy presently available.[19] The oil and natural gas saved today are less polluting than the coal and less hazardous than the nuclear power that the nation will become increasingly dependent upon as oil and gas reserves are further depleted. In a world in which rising energy prices seem inescapable, saving energy is in many cases cheaper than producing more energy to meet demand.

Conservation buys time. The more energy the nation saves now, the longer it postpones the arrival of the time when it will need to rely on new sources of energy and new energy technologies. The longer the country postpones that time, the smoother the transition should be. Furthermore, conservation now reduces the likelihood that the nation will face the hardship and sacrifice that would occur if oil and gas run short before the country is ready to make the transition to new sources and technologies.

How Much Energy Can The Nation Save?

The United States, with 8 percent of the free world's population, consumes 40 percent of the free world's energy and produces 52 percent of the free world's gross national product. Among the major industrial nations of the world, only Canada has higher per capita energy use than the United States. America consumes one-third of the oil used in the world each day, with American motorists accounting for one-ninth of the world's daily oil consumption.[20] How much of this energy is wasted, or to phrase the question in terms of conservation potential, how much energy can the nation save without major lifestyle changes?

Estimates of how much energy the nation can reasonably conserve vary considerably. Dr. Barry Commoner has calculated that, in theory at least, the country could save 85 percent of present consumption and that practicably achievable savings could be as much as 55–60 percent.[21] Denis Hayes of the Worldwatch Institute has argued that the United States could save 50 percent of the energy it now consumes without suffering a decline in the standard of living.[22] President Carter has, on at least one occasion, stated his belief that America wastes about half the energy it uses.[23] Carla Hills, co-chairperson of the board of directors of the Alliance to Save Energy, has asserted that reducing present energy consumption by one-third would not impair the health of the economy.[24] After comparing energy use in Sweden and America, Lee Schipper and Alan J. Lichtenberg have concluded that if this country adopted Swedish energy conservation measures, it could save 30 percent of the energy it consumes.[25]

Predictably, the energy companies have rejected claims that savings of these magnitudes are possible, but they have not offered estimates of their own. Mobil, for example, has undertaken an advertising campaign to debunk "the myth of a wasteful America." According to Mobil, reducing energy consumption by 50 percent would drastically change lifestyles and limit individual freedom of choice.[26]

This difference of opinion over whether the country can reduce its energy use by 50 percent through conscientious conservation alone, as Hayes argues, or only through painful curtailment, as Mobile asserts, need not be resolved. Regardless of which position is more nearly correct, no reasonable person can seriously deny that many opportunities for achieving significant conservation exist today and that seizing these opportunities will benefit the nation as a whole and its citizens individually.

Unfortunately, the country has not even begun to take advantage of many reasonable opportunities to save energy. After energy use fell in 1974 and 1975 as a result of the embargo and the recession it set off, America's appetite for energy began to grow again in 1976. Several circumstances explain why people have spurned conservation opportunities. One of the primary reasons—the invisibility of the current energy problem—was discussed earlier in this chapter. Until people perceive the problem and comprehend that it affects them personally, they have no motivation to take measures for dealing with it.

A second reason that the public has not embraced energy conservation is the deceptive media campaign waged against conservation. Conservation's detractors contend that conservation will result in personal sacrifice, a loss of freedom, and a lower standard of living. Comparative studies of various nations' energy consumption patterns expose the fallacy of the anticonservationists' argument. Sweden produces a slightly higher gross national product per capita than the United States and enjoys a similar standard of living, yet it consumes only 60 percent as much energy as the United States.[27] Among the most important factors contributing to the lower energy consumption in Sweden are the use of smaller cars, greater reliance on mass transit, better insulation and tighter construction of buildings, more efficient industrial processes, and the use of cogeneration and district heating.[28] None of these conservation measures require undue sacrifice or threaten personal freedom, but, more importantly, the Swedish experience refutes the assertion that effective energy-saving efforts will necessarily produce a lower standard of living.

Perhaps the main reason that America has not achieved substantial energy conservation is that energy is underpriced.[29] To anyone who has paid gasoline, home heating oil, electricity, or natural gas bills in the last few years and who remembers how these bills have skyrocketed since the embargo, a statement that energy is underpriced undoubtedly seems ridiculous. To say that energy is underpriced, however, only means that the price consumers pay for energy is lower than the cost of providing energy to replace that which is consumed. The result of the underpricing of energy is that people use more energy than they would if they had to pay prices reflecting replacement costs. The International Energy Agency, to which the United States belongs, has surveyed the energy use of member nations and found that several nations where the cost of energy (prices plus taxes) is higher and, thus, more reflective of replacement costs than it is here have compiled better conservation records than the United States.[30] On the basis of these analyses, the International Energy Agency has concluded that energy prices and taxes are the most important features of energy conservation plans.[31]

How Can Government Encourage Conservation?

Even if energy prices and taxes are the keys to significant energy savings, a comprehensive energy conservation program should not rely on them to the exclusion of other conservation strategies. Other approaches to conservation include exhortation and education, incentives, and regulation.

Exhortation and education efforts are most effective when they are combined with other strategies. The past few years have clearly demonstrated that unless the people believe that an emergency exists or feel a sense of urgency, appeals to the nation to do its patriotic duty (e.g., to engage in the "moral equivalent of war") will have very little impact. Similarly, campaigns to inform the public about the energy problem's impact on inflation and unemployment have resulted in few changes in attitudes or behavior.

Education efforts, however, are necessary to the ultimate success of other strategies. For example, under government mandate, automobile and appliance manufacturers have made, and will continue to make, technological changes to improve the energy efficiency of their products. The energy efficiency labeling of automobiles and appliances, required by EPCA, informs consumers of the improvements and educates them about the energy and financial consequences of the purchasing decisions they are considering.

Incentive strategies include grants, subsidies, tax credits, low interest loans, and government-sponsored research and development. Federal conservation programs have relied heavily on incentives—principally income tax credits, grants, and loans—to encourage people to improve the thermal efficiency of their homes or to install solar energy.[32] States have also enacted incentives—especially income, sales, and property tax credits or exemptions—to promote solar energy.[33] The rationale for these types of incentives is that they lower what for many people are significant barriers to conservation undertakings by reducing high initial capital costs or by alleviating the economic consequences (e.g., property tax increases) of making the improvements. Some people, however, oppose certain incentives, such as the tax credits for home insulation, because they object to the government's paying people to do something that is already in their economic self-interest.

Despite occasional opposition to particular incentives, incentives will generally be very popular politically. For this reason, states and localities should consider including incentives in energy conservation programs as a means of sweetening the bitter taste of tough conservation measures. For example, public transportation fare subsidies may partially defuse opposition to strategies such as congestion pricing or parking bans that are intended to convince people to abandon their cars for less energy-intensive modes of transportation.

Regulation is the traditional government response to a problem, like the energy problem, that requires public guidance or restraint on the private sector. In the energy conservation field, federal regulatory measures have included allocation of energy supplies during shortages; promulgation of energy efficiency standards for appliances, automobiles, and buildings; and prohibitions, such as the requirement in the NEA that new industrial plants and utilities burn fuels other than oil and natural gas. Other books in this series examine regulatory strategies that states and localities have considered or enacted to save energy.

Government also employs its regulatory powers to control energy prices. Although price regulation does insulate consumers from the shock of higher prices, it also creates an erroneous impression that energy remains relatively cheap. As a result, people make fewer energy-saving investments than they would if prices were not artificially suppressed.

Energy prices and taxes will ultimately determine the success or failure of government efforts to encourage greater energy conservation. Modifying or lifting price controls on oil and natural gas to eliminate the distorted signals that controls send consumers would enhance the chances for success, but making the necessary adjustments is a matter of federal, rather than state, responsibility. At the state level, utility commissions can correct another set of erroneous signals by changing utility rate structures to reflect marginal cost pricing or peak period pricing principles.[34]

Although the states have only limited opportunities for affecting energy prices, they have a wide range of fiscal instruments—taxes, fees, civil penalties, and deposits—from which to choose in fashioning energy conservation policies. This book focuses on several of the most promising of these instruments.

NOTES TO CHAPTER 1

1. Bureau of Mines, U.S. Department of the Interior, "U.S. Energy Use Down in 1974 After Two Decades of Increases" (Washington, D.C., April 3, 1975).

2. 15 U.S.C. §§ 751–756.

3. The books in this series discuss many of the proposed and enacted state energy conservation programs. Additional information on state efforts can be obtained from the Energy Project of the National Conference of State Legislatures, located in Denver, Colorado. This project publishes "Energy Report to the States," a biweekly newsletter, and *Energy: The States' Response*, an annual compilation of enacted energy legislation.

4. P. L. 94–163 (EPCA), codified in part at 42 U.S.C. §§ 6201 et seq. and 15 U.S.C. §§ 2001–2012; and P. L. 94–385 (ECPA), codified at 42 U.S.C. §§ 6801 et seq. and 15 U.S.C. §§ 757 et seq.

5. P. L. 94–163, §§ 362–366 codified at 42 U.S.C. §§ 6322–6326.

6. The results of the poll were announced on the CBS News Special Report, "ENERGY: The Facts . . . The Fears . . . The Future," aired August 31, 1977.

7. According to the 1978 Gallup poll, 60 percent of the public knew that the United States imported oil. A year earlier, another Gallup poll found that only 52 percent of the public was aware that the nation imported oil. George Gallup, "Public Cold to Energy 'Crisis' Warnings," *Washington Post*, April 30, 1978, p. A–15.

8. This estimate was announced by President Carter in his address to the nation on energy on November 8, 1977. See *Washington Post*, November 9, 1977, p. A–20 for the prepared text of the address.

9. According to one analysis of the relationship, each 10 percent decline in the value of the dollar overseas results in a 1.5 percent increase in consumer prices. In the period between November 1977 and November 1978, the dollar declined 18 percent. "The Dollar in Trouble," *Washington Post*, November 1, 1978, p. A–14.

10. Carter's address, *supra* note 8.

11. Executive Office of the President, Energy Policy and Planning, *The National Energy Plan* (Washington, D.C.: U.S. Government Printing Office, 1977).

12. The National Energy Act, which President Carter signed into law on November 9, 1978, has five parts: (1) The Public Utility Regulatory Policies Act of 1978, P. L. 95–617, 92 Stat. 3117 (1978); (2) The Energy Tax Act of 1978, P. L. 95–618, 92 Stat. 3174 (1978); (3) The National Energy Conservation Policy Act of 1978, P. L. 95–619, 92 Stat. 3209 (1978); (4) The Power Plant and

Industrial Fuel Use Act of 1978, P. L. 95—620, 92 Stat. 3289 (1978); (5) The Natural Gas Policy Act of 1978, P. L. 95—621, 92 Stat. 3350 (1978).

13. *The National Energy Plan, supra* note 11 at 94.

14. J.P. Smith, "Energy Bills Short of Oil-Cut Goal," *Washington Post*, October 1, 1978, pp. A—1, 4; Office of Public Affairs, U.S. Department of Energy, "The National Energy Act, Press Kit" (Washington, D.C.: November 1978), § III.

15. William Greider, "Array of Experts Disputes 'Energy Crisis' Forecasts," *Washington Post*, July 23, 1978, p. A—8.

16. L. Schipper and J. Darmstadter, "The Logic of Conservation," 80 *Technology Review* 41, 42 (January 1978).

17. *See generally* N. L. Dean, *Energy Efficiency in Industry* (Cambridge, Massachusetts: Ballinger Publishing Co., 1979); R.A. Friedrich, *Energy Conservation for American Agriculture* (Cambridge, Massachusetts: Ballinger Publishing Co., 1978).

18. *See generally* G.P. Thompson, *Building to Save Energy—Legal and Regulatory Approaches* (Cambridge, Massachusetts: Ballinger Publishing Co., 1979).

19. *See* D. Hayes, *Energy: The Case for Conservation*, Worldwatch Paper 4 (Washington, D.C.: Worldwatch Institute, 1976), pp. 20—25; Schipper and Darmstadter, *supra* note 16 at 48.

20. D. Yerkin, "The Wrong Energy Debate," *Washington Post*, January 15, 1977, p. A—23.

21. B. Commoner, *The Poverty of Power* (New York: Alfred A. Knopf, 1976), p. 156.

22. Hayes, *supra* note 19 at 7.

23. Presidential press conference, October 27, 1977.

24. "The Energy Agenda," *Washington Post*, June 5, 1978, p. D—12.

25. L. Schipper and A.J. Lichtenberg, "Efficient Energy Use and Well-Being: The Swedish Example," *Science* 194 (December 3, 1976): 1001, 1012.

26. *Washington Post*, November 6, 1977, p. C—2, the Mobil advertisement entitled "Now more than ever, let's have the facts."

27. Schipper and Lichtenberg, *supra* note 24 at 1001.

28. *Id.* at 1012.

29. A recognition of the need to correct the underpricing of energy is the basis of one of the underlying principles—that "energy prices should generally reflect the true replacement cost of energy"—of President Carter's National Energy Plan. *See The National Energy Plan, supra* note 11 at 29—30. *See also* C.J. Hitch, "Energy in Our Future," 43 *The Key Reporter* 2, 3 (Summer 1978).

30. Organisation for Economic Co-operation and Development, *Energy Conservation in the International Energy Agency: 1976 Review* (Paris, 1976).

31. *Id.* at 7.

32. *E.g.*, ECPA authorizes a $200 million grant program to permit low income persons to weatherize existing homes and a $200 million demonstration program to identify incentives to encourage people to invest in conservation-related home improvements; the NEA authorizes a host of measures, including residential conservation tax credits, solar and wind energy tax credits, conservation loans of up to $2,500 for homeowners and renters with incomes below

their area's median family income, an extension of ECPA's program of weatherization grants for low income families, a weatherization grant program administered by the Farmers Home Administration for low income families in rural areas, and federally assisted loans of up to $8,000 to assist homeowners, regardless of income, to purchase and install solar energy equipment.

33. In 1977 and 1978, twenty-six states enacted thirty-six laws authorizing solar or alternative energy tax credits or exemptions. These laws are collected and printed in *Energy: The States' Response—1977* and *Energy: The States' Response—1978*, publications of the Energy Policy Project of the National Conference of State Legislatures, Denver, Colorado.

34. *See* F.J. Wells, *Utility Pricing and Planning—An Economic Analysis* (Cambridge, Massachusetts: Ballinger Publishing Co., 1979).

✳ *Chapter 2*

Advantages and Disadvantages
of Economic Disincentives

The economic disincentives discussed in this book are monetary charges levied by the government for the principal purpose of discouraging energy-wasteful conduct.[1] These charges include a variety of fiscal instruments, such as taxes, regulatory fees, deposits, and civil penalties. Some of these disincentives generate substantial revenue, others do not; some promote conservation generally by raising the cost of energy, while others have a more specific aim, such as persuading people to choose a more energy-efficient machine or technology or to shift to a less energy-intensive method of operation. Despite their differences, economic disincentives share the common goal of encouraging energy conservation.

Regardless of their potential, economic disincentives are not a panacea for the nation's energy problem. Disincentives may not work at all in some circumstances. In other situations, a regulatory or incentive strategy may be more effective than a disincentive measure. Moreover, some disincentives may have undesirable side effects that create problems as serious as those they are designed to alleviate.

Recognizing that economic disincentives have flaws and shortcomings should not remove them from further consideration as a means of dealing with the energy problem. No strategy, whether it involves a regulatory mechanism, an incentive, or a disincentive, is flawless. The question is whether the pluses of a particular strategy exceed its minuses. In general, the advantages of economic disincentives seem to outweigh their disadvantages. This chapter discusses the major general advantages and disadvantages of economic disincentives, so that policymakers can conduct the weighing process for themselves. The discussion concludes with an analysis of the political obstacles that disincentives will have to overcome.

ADVANTAGES OF ECONOMIC DISINCENTIVES

Economic disincentives will capture people's attention and focus it on the energy problem more dramatically than educational efforts or regulatory programs. Disincentives will wake the nation up to the need to develop a conservation consciousness and will make individuals and businesses reassess their energy consumption habits.

Every day people conduct their lives unaware that activities as seemingly innocuous as buying soft drinks and beer contribute to the nation's energy problem. Several years ago, most beverage containers were glass and were designed to be returned for reuse. More recently, however, container manufacturers and bottlers have catered to the consumer's desire for convenience by providing nonreturnable bottles and cans. Unfortunately, accommodating consumer convenience has had adverse environmental (solid waste disposal) and energy consequences. It takes 78 percent more energy to make an aluminum can from virgin materials than from secondary materials (e.g., recycled cans).[2] The energy required to make a single glass bottle and reuse it ten times is less than one-third of what is needed to produce ten nonrefillable bottles of the same size.[3]

Beverage container deposit laws ("bottle bills"), which seven states and several localities have enacted,[4] would alert the public to the connection between soft drinks and beer and the energy problem. The refundable deposit (usually 5 cents) required by these laws is intended to eliminate nonrefillable bottles and cans and to encourage the reuse or recycling of all beverage containers. According to several studies, bottle bills will significantly cut waste (garbage and energy) at the expense of a small reduction in consumer convenience.[5]

Automobiles have a much more obvious connection to the energy problem, yet many people apparently find it easy to ignore window stickers providing fuel economy and annual fuel cost information. Economic disincentives, however, such as the program of automobile efficiency taxes and rebates discussed in Chapter 4, would command a purchaser's attention. Even the automobile companies cannot quarrel with this premise. During the recession in 1974 and 1975, the auto manufacturers resorted to rebates to encourage sales of slow-selling small cars. In December 1978, they raised the prices of their less efficient big cars because better than expected sales of these models were jeopardizing the companies' chances of meeting the federally mandated sales-weighted fleet fuel economy standard for the 1979 model year.

If experiences in the past decade with pollution control programs are any indication, regulatory efforts to achieve energy conservation may draw the public's notice only insofar as is necessary to devise ways to avoid complying with regulatory requirements. One type of disincentive, an economic civil penalties program (see Chapter 3), could be effective in shifting the focus from avoidance to compliance. A state, for example, could enact lighting efficiency standards that mandate the retrofitting of existing buildings. Some building owners may ignore the standards if retrofitting is expensive and if the savings from not retrofitting seem likely to exceed any fines that would be levied for noncompliance. As a key element of conservation standards, a civil penalties provision authorizing the monitoring agency to collect a penalty based on the savings realized by the owner as a result of his noncompliance would remove the incentive to delay compliance. Faced with the possibility of having to pay this type of penalty, individuals and businesses subject to energy conservation regulations or standards would reassess their energy consumption decisions.

A second major advantage of economic disincentives is that they permit people to reduce their energy use in the way that maximizes personal satisfaction. In contrast to conservation strategies that consist of government edicts that everyone is expected to obey, disincentives give individuals freedom to choose when, how, and even whether to save energy. Disincentives offer people the option of undertaking conservation efforts and saving money or maintaining consumption habits and spending less income on other things. If people decide to reduce their energy consumption, disincentives allow them to establish schedules that reflect their individual circumstances. For instance, in response to stiff annual automobile registration fees for big cars, some gas-guzzler owners would trade cars immediately, while others could postpone trading cars until the time was right for them (e.g., when their children had grown up). Disincentives also preserve individual choice over how to cut energy use. Gasoline taxes, for example, may result in some people cutting their pleasure driving so that they can continue driving to work and in others commuting by public transportation so that they can maintain their pleasure driving.

A third major advantage of disincentives is that, in certain cases, they may be the only constitutionally permissible means that a state has for effectively dealing with a particular facet of the energy problem. For instance, strategies for attacking the problem of gas-guzzling automobiles include banning their manufacture, banning their importation into and sale in a state, and imposing tough fuel economy standards. State legislation to implement any of these approaches would almost surely be invalidated by the courts on the grounds that the legislation imposed an undue burden on interstate commerce or was preempted by federal legislation. On the other hand, a program of state automobile efficiency taxes and rebates would probably survive similar challenges (see Appendix 4–1).

A fourth major advantage of disincentives is that they can serve to correct market failures or distortions. Chapters 1 and 3 discuss the fact that energy is regarded as underpriced because energy prices do not reflect true replacement costs. Many experts consider the underpricing of energy the most serious impediment to efforts to encourage energy conservation. Disincentives give states a means with which to remedy the distorted price signals energy users currently receive.

In the process of removing the obstacle to conservation presented by the underpricing of energy, many economic disincentives will produce substantial revenue. The revenue-raising potential of disincentives is their final, and perhaps most politically popular, advantage. Unlike the lifting of energy price controls, another approach to correcting the underpricing problem, disincentives will generate revenues for the state rather than higher profits for the energy companies. States can spend these revenues in ways that neutralize either general disadvantages of disincentives or opposition to particular disincentives. For example, many people fear that disincentives will have an adverse impact on the economy. States could minimize the impact by recycling the revenues from disincentives back into the economy. More specifically, one of the major disadvantages of disincentives is the disproportionately heavy burden they may impose on the poor and the elderly. Disincentive revenues provide a means to alleviate this burden. States can also use disincentive revenues to provide less energy-intensive alternatives to existing practices. The objections of the motoring public to automobile disincentives, for example, may be quieted if the revenues from these measures are devoted to financing transportation alternatives to the automobile. Emphasizing the benefits that disincentive revenues can provide will be one of the key tactics in any campaign to have disincentives enacted.

DISADVANTAGES OF ECONOMIC DISINCENTIVES

The preceding recitation of the advantages of economic disincentives is not intended to create the impression that disincentives are a cure-all for the energy problem that ails the nation. Like any new and strong medicine, disincentives have limitations and a potential for producing undesirable side effects. As parents who have given their children cod liver oil know, advocates of disincentives will discover that efforts to administer disincentives will encounter fierce resistance.

Disincentive opponents will center their campaigns to defeat disincentive proposals around the strategy's disadvantages. Proponents can steal some of their opponents' ammunition by designing disincentives to eliminate potential adverse impacts. Some of the disadvantages of disincentives, however, involve political obstacles that proponents will find difficult to overcome.

To rebut assertions that economic disincentives will capture the public's attention and encourage people to reduce their energy use, opponents will probably draw an example from recent history. Beginning in 1976, gasoline consumption has risen to record levels every year even though prices have also risen. This trend causes one to wonder whether the public, given the freedom of choice that disincentives allow, will elect to pay the disincentives rather than make the adjustments that will save energy (and money).

Instead of demonstrating that all disincentives will be ineffective in encouraging conservation, the gasoline situation simply illustrates one of the limitations of disincentives. High gasoline prices have not curbed gasoline demand in part because the short run demand for gasoline is inelastic (i.e., not very responsive to price increases).[6] In any circumstance in which the demand for energy is inelastic, economic disincentives will not be particularly effective in encouraging people to reduce their energy consumption.

Even in a situation involving inelastic demand, however, disincentives may be justified. Several factors contribute to the inelastic demand for gasoline, but perhaps the most important is that the vast majority of people have no choice but to continue driving if they want to maintain their mobility. Higher gasoline taxes may not cut gasoline demand in the short run, but even a small tax increase would produce significant revenues. This money could be spent to provide or improve transportation alternatives to the automobile. Once people have a choice, disincentives will probably be more effective in reducing energy consumption.[7]

A second major disadvantage of disincentives is their potentially adverse economic impact. Disincentive opponents will probably argue that this conservation strategy will hurt the economy by reducing individuals' purchasing power, aggravating inflation, and slowing economic growth. The basic issue, though, is whether the proposed remedy—disincentives—is more harmful than the problem it addresses.

Chapter 1 highlights the link between the problem—the nation's inability to curb its appetite for foreign oil—and the state of the domestic economy. In 1977, the nation imported 47 percent of its oil at a cost of $42.4 billion. In 1978, as oil production from Alaska's North Slope reduced imports to 43 percent of total demand, the bill for imported oil declined slightly to $39.5 billion. These outlays for foreign oil have contributed to record trade deficits of $26.6 billion in 1977 and $28.5 billion in 1978. Soaring trade deficits have shaken the world money markets' confidence in the soundness of the dollar, and as the value of the dollar has dropped abroad, inflation has risen at home. The decline in the value of the dollar between mid-1977 and the end of 1978 has added over 1 percent to the domestic inflation rate. In contrast, President Carter's National Energy Plan, which proposed relying on several disincentives to reduce oil imports, would have raised the inflation rate only 0.4 percent.[8]

The decision in December 1978 of the Organization of Petroleum Exporting Countries (OPEC) to boost their prices in 1979 by 14.5 percent dealt another blow to the American economy. This price hike will increase the nation's bill for imported oil by $4 billion and will add $2 billion to the country's trade deficit.[9] Gasoline and home heating oil prices will rise at least 3 cents per gallon. The price increase will slow economic growth by 0.1 percent and add 0.2-0.4 percentage points to the 8 percent inflation rate economists have predicted for 1979.

Determining whether the economic benefits of a particular disincentive will exceed its economic costs is beyond the scope of this book. States and localities should, however, thoroughly analyze the impact that proposed disincentives will have on prevailing economic conditions before enacting any disincentives. To increase disincentives' chances of surviving economic impact analyses, proponents can design these measures in ways that minimize the possibility of their having adverse economic consequences. First, stiff disincentives should be proposed only when there is a reasonable (and less energy-consuming) alternative to the product or practice on which the disincentive will be imposed. As long as a realistic alternative exists, people will be able to protect their purchasing power by turning to the alternative to avoid paying the disincentive. Second, in cases where there are no adequate alternatives, disincentives should be phased in gradually as alternatives are developed. Gradually imposing a disincentive will also cushion the economic shock from a substantial tax or fee, but as Chapter 3 warns, phasing in a disincentive could dilute its effectiveness. Finally, the adverse economic impact of disincentives can be reduced through a recycling of the revenues they generate. Disincentives can finance tax cuts, complementary rebates (e.g., automobile efficiency rebates), or energy conservation projects (e.g., transit or paratransit alternatives to the automobile and energy-saving home improvements).

Not only should legislators worry about the economic impact of disincentives, they should also be concerned with the measures' social impacts, such as the effect on the disadvantaged. Their potential for imposing a heavy burden on the poor and the elderly is the third major disadvantage for disincentives. Many of these persons already have trouble paying their energy bills, and disincentives, if not carefully designed, may aggravate their problems. From the perspective of the poor, the freedom of choice allowed by disincentives may only be a cruel illusion. Whether they make the changes the disincentives are intended to encourage or pay the disincentive, they may end up, at least for the time being, poorer.

The harshness of this result can be alleviated by both the people and the government. The poor can help themselves by taking advantage of incentives that are coupled with the disincentives. For example, by purchasing a fuel-efficient car, a low income family could enjoy not only an immediate cash rebate, but also the long term savings of lower gasoline bills. In many cases, however, the government will have to provide relief from the impact of disincentives. Chapter 5, for example, examines several measures that states can either use now to ease the burden of rising residential costs or enact in conjunction with disincentives imposed in the future on residential energy services to cushion their impact on the poor.

Designing disincentives to avoid adverse economic and social side effects will probably not significantly dampen the groundswell of political opposition that will greet their proposal. Many powerful interest groups will undoubtedly devote their resources to defeating disincentive legislation. The auto workers, for example, opposed the National Energy Plan's proposal for gas guzzler taxes and efficiency rebates, and it is conceivable that they will mount campaigns against state automobile efficiency tax-rebate legislation. Soft drink and beer bottlers and beverage container manufacturers have fought beverage container deposits wherever bottle bills have surfaced, and although they have lost some battles, many of their campaigns have been successful.[10]

Perhaps the loudest outcries against economic discincentives will come from the people who will have to pay them. Taxpayers have always had an understandable aversion to higher taxes, and politicians are becoming increasingly cautious about antagonizing voters by increasing their tax burden. In the aftermath of the tax revolt ignited by Proposition 13 in California, the chances for enactment of economic disincentives must seem particularly slim to a casual observer. A closer examination of the lessons of Proposition 13, however, rekindles a degree of hope for disincentives.

Proposition 13 and its progeny carried a message more complex than a simple reaffirmation of the public's dislike of high taxes. These tax reduction initiatives signaled the public's dissatisfaction with the quality of the services supported by their taxes. By voting for these measures, people communicated their disenchantment not only with welfare rolls that continue to grow but also with schools that fail to teach their children reading, writing, or arithmetic. The tax revolt, therefore, is essentially a protest by consumers who feel they are not receiving full value for their taxes.

Proponents of disincentives can capitalize on these feelings of frustration by showing how disincentive revenues can provide or improve needed services (e.g., public transportation). If the public perceives that disincentives will have positive benefits, they may be more favorably disposed to their enactment. In Houston, for example, voters passed a referendum in August 1978 that created a transit authority and added 1 cent to the sales tax to finance the development of an adequate bus system.[11] In this instance, the need for transportation alternatives to the automobile, which became obvious as rush hours began to last longer, overcame Texans' particularly strong dislike of higher taxes.

Unfortunately, the task of selling economic disincentives to the American people will be more formidable than the task faced by the referendum backers in Houston. At least in Houston everyone knew a serious problem existed because they were caught in the middle of it twice a day. To sell disincentives, proponents will first have to convince the public and state and local legislators and officials that there is an energy problem and that states and localities can deal with it by encouraging energy conservation. Next, they will have to publicize the benefits of conservation for individuals and society as a whole in order to counter the false impression that conservation will result in hardship or deprivation. Finally, proponents will have to show that the advantages of employing economic disincentives to encourage conservation outweigh the disadvantages. Even if proponents make a convincing sales pitch, however, it is likely that many people will not buy disincentives until the nation's energy problem becomes much more conspicuous.

NOTES TO CHAPTER 2

1. *See* W.A. Irwin and R.A. Liroff, *Economic Disincentives for Pollution Control: Legal, Political and Administrative Dimensions*, EPA−600/5−74−026, prepared for the U.S. Environmental Protection Agency by the Environmental Law Institute (Washington, D.C.: U.S. Government Printing Office, 1974). Irwin and Liroff define disincentive as "a monetary charge levied by government on conduct which is not illegal but which does impose social costs, for the principal purpose of discouraging the conduct" (*id.* at 6).

The definition of disincentives used in this book is broader than Irwin and Liroff's, so that the term can be used to characterize the beverage container deposits discussed in this chapter, the economic civil penalties discussed in Chapter 3, and the taxes and fees analyzed in Chapter 4.

2. U.S. Environmental Protection Agency, *Fourth Report to Congress on Resource Recovery and Waste Reduction* (Washington, D.C.: U.S. Government Printing Office, 1977), p. 69.

3. *Id.*

4. Connecticut, Delaware, Iowa, Maine, Michigan, Oregon, and Vermont have beverage container deposit laws. Among the localities that have bottle bills are Berkeley, Mill Valley, Tiburon, and Marin County, California; Highland Park, Illinois; Bowie and Montgomery County, Maryland; New Hope, Minnesota; Columbia, Missouri; Oberlin, Ohio; and Fairfax and Loudon counties, Virginia. Several local bottle bills have been invalidated by the courts (e.g., Ann Arbor, Michigan, and Cayuga County, New York) and one (Howard County, Maryland) was repealed by referendum in 1978 two years after the voters had approved it.

5. For a more complete discussion of the energy and economic impacts of beverage container deposits, *see* N.L. Dean, *Energy Efficiency in Industry* (Cambridge, Massachusetts: Ballinger Publishing Co., 1979); EPA, *supra* note 2 at 67−75.

6. The gasoline tax section of Chapter 4 discusses several studies that have found the short run demand for gasoline to be inelastic.

7. Estimates of the elasticity of demand for gasoline, discussed in Chapter 4, suggest that demand will be more elastic in the long run (i.e., higher gasoline taxes will be more effective in reducing gasoline consumption in the long run).

8, The chairman of the Federal Reserve Board, G. William Miller, announced this estimate of the inflationary impact of the National Energy Plan in congressional testimony in March 1978. James L. Rowe, Jr., "Cuts in Oil Imports Asked," *Washington Post*, March 16, 1978, p. C−1.

9. The estimates in this paragraph of the economic impact of the OPEC 14.5 percent price increase are the Carter Administration's projections. Art Pine, "Oil Price Increase to Weaken '79 U.S. Economy Somewhat," *Washington Post*,

December 18, 1978, pp. A–1, 6; and "The Aftermath of OPEC's Shock" *Business Week*, January 8, 1979, p. 14.

10. *E.g.*, in Washington State in 1970 the beverage container coalition defeated Initiative 256, which would have required a 5 cent deposit on soft drink and beer containers sold for consumption in the state. For an account of the campaign to defeat Initiative 256, see W.H. Rodgers, Jr., *Corporate Country* (Emmaus, Pennsylvania: Rodale Press, 1973), pp. 3–27. In 1976 the coalition financed a successful campaign against a container deposit initiative in Colorado.

11. Bill Curry, " 'Auto-Mated' Houston to Vote on Taxes vs. Congestion," *Washington Post*, August 11, 1978, p. A–3; and Bill Curry, "Houston, Bucking National Trend, Votes Tax Increase," *Washington Post*, August 14, 1978, p. A–5.

✳ *Chapter 3*

Legal and Design Considerations
for Economic Disincentives

As the preceding two chapters have indicated, some serious barriers lie in the way of economic disincentives being accepted by the people or enacted by their representatives. Geopolitical events (e.g., a disruption of the world oil market such as that caused by the political turmoil in Iran) or a decision by government officials to exercise bold leadership, however, could neutralize these barriers. In anticipation of a political climate more favorable to economic disincentives, proponents of disincentives should be prepared with proposals free from legal or policy problems. This chapter, therefore, examines some of the legal and design considerations that should be taken into account in drafting economic disincentives.

LEGAL ASPECTS OF
ECONOMIC DISINCENTIVES

Economic disincentives intended to encourage energy conservation do not fit snugly into any of the legal system's traditional cubbyholes. Unlike most taxes, economic disincentives have very important regulatory objectives. Unlike most regulatory programs, however, many disincentives have the potential for producing substantial revenue. In several respects, economic disincentives are analogous to environmental charges, a variety of strategies that reflect an economics-based approach to environmental improvement.[1] Despite the relative novelty of these strategies, it appears, based on the experience to date, that carefully designed economic disincentives should survive legal challenges.[2]

This section briefly examines the scope of the powers under which states and localities can enact disincentives and the major federal constitutional limitations on the exercise of those powers. State constitutions and laws, municipal charters and ordinances, and state court decisions may impose additional limitations on the powers of the states and localities to adopt disincentive strategies. A complete analysis of these sources for possible additional limitations, however, exceeds the scope of this book. This section, therefore, represents only a starting point. It is intended to serve only as a checklist of major issues; it cannot serve as a substitute for the in-depth legal analysis that is necessary for designing economic disincentives that avoid all legal pitfalls.

State Powers and Economic Disincentives

Under the tenth amendment of the U.S. Constitution the sovereign states and the people possess all "powers not delegated to the United States by the Constitution nor prohibited by it to the States." Either the states' police powers or their taxing powers will provide the basis for the enactment of economic disincentives.

The police power is the power of the state to regulate the private affairs of its citizens to further important public interests. The usual shorthand definition of the police power describes it as the power to legislate to protect the public health, safety, welfare, and morality. Even this definition of the power is perhaps too narrow because the police power "embraces an almost infinite variety of subjects."[3] According to Justice Oliver Wendell Holmes, "the police power extends to all the great public needs. . . . It may be put forth in aid of what is sanctioned by usage, or held by the prevailing morality or strong and preponderant opinion to be greatly and immediately necessary to the public welfare."[4]

The breadth of the police power is a reflection as well as a product of the power's dynamic character. The concepts of public health, safety, welfare, and morality are fluid, capable of adapting to the tenor of the times. Justice Holmes recognized this when he wrote that "circumstances may so change in time or so differ in space as to clothe with . . . a [public] interest what at other times or in other places would be a matter of purely private concern."[5]

Although the police power has a broad and ever-changing reach, the courts have developed standards for determining whether, in a particular instance, a state has overstepped its authority. The general standard is the familiar, but amorphous, one of "reasonableness." In

1894, the Supreme Court formulated what has become the classic statement of this standard: "To justify the State in . . . interposing its authority in behalf of the public, it must appear, first, that the interests of the public . . . require such interference; and, second, that the means are reasonably necessary for the accomplishment of the purpose, and not unduly oppressive upon individuals."[6] In affirming the continuing validity of this statement, the Court in *Goldblatt v. Town of Hempstead* noted that "[e]ven this rule is not applied with strict precision, for this Court has often said that 'debatable questions as to reasonableness are not for the courts but for the legislature. . . .' "[7]

Controversial or novel legislation enacted under the police power has often provoked litigation in which the reasonableness standard is applied. Anticipating such litigation, a legislature can fortify legislation to improve its chances of withstanding judicial scrutiny. In addition to avoiding obvious deficiencies (e.g., vagueness or insidiously discriminatory provisions), the legislature should set forth a detailed justification for its action in a findings and purpose (or preamble) clause. In general, the findings should describe the problem that the legislation addresses and the way in which the problem affects the public health, safety, welfare, or morals. In the case of energy conservation legislation, drafting findings that describe the energy problem and its adverse impacts on the public should be fairly simple. The legislature, in most instances, should also include in the preamble an explanation of why it chose the particular means for dealing with the situation. For example, the findings could detail the failure of less drastic alternatives or explain that the means chosen are the most effective of the alternatives. A lengthy findings and purpose clause will not guarantee that courts will uphold a state's exercise of its police power. A legislative statement, however, that clearly demonstrates the need for government action to protect the interests of the public should increase the likelihood that courts will heed the Supreme Court's reminder that "debatable questions as to reasonableness are not for the courts but for the legislature." A deference to legislative judgment concerning means is particularly important when the means are as novel as economic disincentives.

In a notable case involving the police power and a measure that fits the definition of economic disincentives, the Oregon Court of Appeals deferred to the state legislature's judgment when it held that the beverage container deposit legislation ("bottle bill") was "unquestionably a legitimate exercise of the police power."[8] After recognizing that "[s]election of a reasonable means to accomplish a state purpose is clearly a legislative, not a judicial, function" and that "the courts may not invalidate legislation . . . because additional and complementary means of accomplishing the same goal may exist," the court discussed the extent of the legislature's discretion in selecting means:

> The legislature may look to its imagination rather than to traditional methods . . . to develop suitable means of dealing with state problems, even though their methods may be unique. Each state is a laboratory for innovation and experimentation in a healthy federal system. What fails may be abandoned and what succeeds may be emulated by other states.[9]

This statement suggests that, like the police power itself, which expands its reach as the times require, the set of tools from which the legislature can choose in exercising its police power may not be limited to traditional ones (e.g., regulatory schemes). This notion raises the interesting question of whether a state can use its police power to enact economic disincentives that, although they are principally intended to discourage energy-wasteful conduct, generate substantial revenue. An analysis of this question requires an examination of the difference between taxes and police power fees.

States collect money from their citizens by imposing taxes under their taxing power and regulatory fees and other charges under their police power. In deciding whether a particular imposition is a tax or a police power exaction, courts have traditionally drawn a distinction based upon the intent of the legislature as to the purpose of the levy. If revenue is the principal purpose of the imposition and its regulatory effects are incidental, it is a tax; if regulation is the principal purpose, the imposition is not a tax even though it "incidentally" produces revenue.[10] Applying this test strictly, courts would probably characterize economic disincentives as regulatory fees rather than taxes.

Because their principal purpose is not the production of revenue, regulatory fees are subject to a legal restriction on their amount. In most instances, these fees must be limited to an amount reasonably related to the cost of administering the regulatory program of which they are a part.[11] In general, if a regulatory fee generates revenue greatly in excess of the costs of regulation, courts will declare the fee to be either a tax or an illegal exercise of the police power.[12] Applying this rule to economic disincentives would, in many if not all cases, produce fees too low to accomplish their principal purpose.

Some courts, however, have recognized exceptions to the rule limiting the amount of regulatory fees. These exceptions may allow states to impose economic disincentives, at effective rates, under their police power. One exception allows the amount of the fee to reflect not only the costs of direct regulation, but also the costs the public may have to bear as a result of the activity that is being regulated.[13] A second exception permits a fee to generate excess revenue if the revenues are used to deal with a problem related to that for which the fee is imposed.[14] A third exception, recognized by many courts in cases involving an activity that a state could prohibit under its police power, allows the state to restrict or prohibit the activity through the imposition of a very high fee.[15] Whether a particular state recognizes or will recognize any of these exceptions can only be determined through an analysis of that state's case law. In the final analysis, however, most states will probably not need to try to fit revenue-producing economic disincentives into one of these exceptions. If none of the exceptions apply to a particular disincentive or if the courts look unfavorably upon a legislature's relying upon "its imagination rather than . . . traditional methods . . . to develop suitable means of dealing with state problems," a state should be able to use the tax power to enact revenue-producing disincentives.

As a sovereign, each state has the inherent power to impose taxes for the purpose of raising revenue to defray the expenses of government. As the Supreme Court has stated:

> The States have a very wide discretion in the laying of their taxes. When dealing with their proper domestic concerns, and not trenching upon the prerogatives of the National Government or violating the guaranties of the Federal Constitution, the States have the attribute of sovereign powers in devising their fiscal systems to ensure revenue and foster their local interests.[16]

Many, though not all, economic disincentives will generate substantial revenue. The key to having these disincentives upheld as taxes is careful legislative drafting. Because opponents may argue that disincentives enacted under the taxing power are really regulatory fees rather than taxes, the legislature should draft the findings and purpose clause of disincentive legislation to emphasize the revenue-raising purposes of the measures. The regulatory objectives (i.e., discouraging energy-wasteful conduct) should receive, at most, a brief mention. In addition to its legal advantages, this approach of focusing on the revenue-raising purposes of disincentives may enhance their political popularity.

Careful legislative drafting, however, cannot hide the regulatory objectives of disincentives and will not deter litigation challenging disincentives. The courts will face the issue of whether economic disincentives, which on their face purport to be taxes, should be struck down because they are in reality regulatory fees producing excessive revenue.

Language in several U.S. Supreme Court cases suggests that economic disincentives will be sustained as taxes. In a case involving a federal tax on firearms dealers, the Court stated that "[e]very tax is in some measure regulatory. . . . But a tax is not any the less a tax because it has a regulatory effect."[17] In response to a challenge to a federal marijuana tax, the Court expanded upon this principle: "It is beyond serious question that a tax does not cease to be valid merely because it regulates, discourages, or even definitely deters the activities taxed. . . . The principle applies even though the revenue obtained is obviously negligible, . . . or *the revenue purpose of the act be secondary, . . .* " (emphasis added).[18]

Although these cases involve taxes enacted by Congress under its tax power, the Supreme Court has applied these principles when reviewing state taxes challenged because of their regulatory effect. In upholding a Washington State excise tax on butter substitutes, the Court stated that "[c]ollateral purposes or motives of a legislature in levying a tax of a kind within the reach of its lawful power are matters beyond the scope of judicial inquiry. . . . "[19] In a 1974 case involving a tax that could serve as an economic disincentive, the Court upheld Pittsburgh's parking tax.[20] The Court noted that this tax raised significant revenues, but that "even if the revenue collected had been insubstantial, . . . or *the revenue purpose only secondary, . . .* we would not necessarily treat this exaction as anything but a tax entitled to the presumption of the validity accorded other taxes imposed by a State" (emphasis added).[21]

Whether the courts of each state currently follow these principles in reviewing taxes that have regulatory effects or whether they will apply them in reviewing economic disincentives is a question that this book cannot answer. At least one state court, however, has upheld a tax whose regulatory objectives were at least as important as its revenue goals. In 1974 the Vermont Supreme Court rejected a constitutional challenge to the state's recently enacted Land Gains Tax.[22] Although the stated purpose of the tax was to provide revenue for property tax relief, the court determined, after analyzing the tax structure, that the legislature could also have intended the tax to deter land speculation. In holding that the legislature acted within its constitutional powers in restricting land speculation by means of the land gains tax structure, the court noted that "[i]t is now beyond question that the legislature may legislate to achieve particular social and economic ends by the manner in which a tax is imposed. . . . "[23] If other state courts adopt this position and follow the principles the U.S. Supreme Court has announced concerning taxes that have regulatory effects, revenue-producing economic disincentives should have little difficulty in being sustained as impositions properly enacted under a state's taxing power.

Local Powers and Economic Disincentives

Before considering whether to enact economic disincentives such as bottle bills, congestion pricing, or parking taxes, local governments must initially determine whether they possess the power to enact them. Local governments receive their express governing authority from the state.[24] Generally, the state confers powers on localities in one of three forms: (1) direct delegation of authority guaranteed in the state constitution (constitutional home rule); (2) delegation through statutory grants of home rule authority (legislative home rule); and (3) express delegation of authority to a particular entity or class (usually, charter rule). In each case, the delegation of authority is often supplemented by state statutes granting additional powers. According to Dillon's Rule,[25] local governments can exercise, in addition to their expressly granted powers, powers necessarily implied or necessarily incident to the expressly granted powers and powers absolutely essential to the declared objects and purposes of the local government. The powers of local governments are subject, however, not only to the federal constitutional limitations that apply to state powers, but also to state constitutional and statutory limitations.[26]

Each local government should examine the source of its powers and the state limitations on those powers to determine whether the police and taxing powers it possesses will support legislation imposing economic disincentives. Generally, states delegate police power to localities in broad terms that make it difficult to sustain a claim that the local government lacked police power authority to act.[27] In most cases, local police power enactments will be challenged on the same grounds as state police power enactments: that the law represents an improper exercise of the power or that it violates one or more provisions of the federal and/or state constitution.[28]

A local government may not impose taxes unless it has been expressly granted the power to tax. Delegations of taxing authority may be either general or specific (i.e., limited to a particular tax). In determining whether a delegation of taxing power permits a certain tax, courts will construe the delegation strictly.[29] Only one of the economic disincentives discussed in this book, parking taxes, is proposed as a measure for local governments to enact under their taxing power. Localities, therefore, should determine whether they can enact parking taxes under existing delegations of taxing authority.

Major Federal Constitutional Limitations on State and Local Powers

Courageous states and localities that decide to serve as "laboratories" for experimenting with economic disincentives must conduct their experiments within the limits of the U.S. Constitution. Even if states and localities scrupulously seek to observe these limits, economic disincentives, because they are novel and controversial, will probably be attacked on constitutional grounds. The major federal constitutional challenges to state and local economic disincentives will be based on the supremacy clause (the preemption doctrine), the commerce clause (the burden on interstate commerce doctrine), and the fourteenth amendment (the guarantees of due process and equal protection). At least for the economic disincentives discussed in this book, these limitations do not present insurmountable obstacles.

The Supremacy Clause. Article VI, Clause 2, of the U.S. Constitution provides:

> The Constitution, and the Laws of the United States which shall be made in Pursuance thereof; and all Treaties made, or which shall be made, under the Authority of the United States, shall be the supreme Law of the Land; and the Judges in every State shall be bound thereby, any Thing in the Constitution or Laws of any State to the Contrary notwithstanding.

From this provision has sprung the doctrine of preemption. This doctrine is applied to resolve conflicts that arise when both a state and the federal government have acted in the same field. Unfortunately, no crystal clear formulation of the doctrine exists; sorting through its various applications can be confusing.

If a state law imposes requirements that so directly conflict with the requirements of federal law that it is impossible to comply with both sets of requirements, the state law will be struck down.[30] Similarly, courts will void state laws that, in purpose or operation, directly conflict with or hinder federal policy.[31] In addition, state laws must yield when Congress expressly declares its intention to exert exclusive authority over the subject matter in question.[32] Cases involving direct conflicts or express preemption are relatively simple compared to those involving claims of implied preemption.

Even though Congress may not have expressly stated its intention to prohibit concurrent or supplementary state regulation of a subject matter, the courts may find that Congress has, by implication, preempted state regulation in that area. In deciding whether federal preemption is implied in a certain field, courts examine the legislation for a number of factors. Courts may find an implied intent to preempt in: (1) the nature of the statute as amplified by the legislative history,[33] (2) the nature of the subject matter regulated if it is one that demands the national uniformity that exclusive federal regulation would provide, [34] (3) the dominance of the federal interest in the subject matter, [35] and (4) the pervasiveness of the federal scheme.[36]

If Congress has neither expressly nor implicitly demonstrated its intent to preempt state regulation in a field, state law may "fill in the gaps" in the federal regulatory scheme.[37] Moreover, several recent cases suggest that even where federal and state laws address the same subject, courts will attempt to harmonize the laws before declaring the state law preempted.[38] The increased willingness to harmonize federal and state law seems particularly strong when the federal law provides for complementary state regulation.[39]

The major federal energy laws—the Energy Policy and Conservation Act (EPCA), the Energy Conservation and Production Act (ECPA), and the National Energy Act (NEA)—do not collectively or individually, expressly or implicitly exhibit a congressional intention to occupy the field of energy conservation to the exclusion of any state regulation. EPCA does, however, expressly preempt certain state laws concerning automobile fuel economy standards and labeling.[40] EPCA, as amended by the NEA, also prohibits certain state laws relating to energy efficiency standards and labeling for household appliances.[41]

In contrast to these express preemption provisions, EPCA extends a generous invitation to the states to participate in the energy conservation effort.[42] The invitation, reaffirmed in the NEA,[43] promises states federal assistance in preparing and implementing state energy conservation plans. EPCA requires the plans to contain certain programs to be eligible for the assistance.[44] States may also include in their plans (1) restrictions on the hours and operations of public buildings, (2) restrictions on the use of decorative and nonessential lighting, (3) transportation controls, (4) public education programs, and (5) *"any other appropriate method or program to conserve and improve efficiency in the use of energy"* (emphasis added).[45]

The state energy conservation plan provisions of EPCA indicate a strong congressional intent that the states play a major role in energy conservation efforts, and they create a strong presumption against preemption challenges to most state energy conservation legislation. With the possible exception of state automobile.efficiency taxes and rebates (discussed in Chapter 4), none of the economic disincentives discussed in this book face serious preemption problems.

The Commerce Clause. The U.S. Constitution lists among the enumerated powers of Congress the power "[t]o regulate Commerce . . . among the several States. . . ."[46] This grant of power to the Congress reflects a recognition by the framers of the importance to the nation of having a free flow of commerce among the states. Over the years, Congress and the U.S. Supreme Court have expanded the concept of interstate commerce to support and justify federal regulation of a variety of private activities.[47]

The commerce clause, as interpreted by the courts, not only provides a broad base of federal power, but also serves as a limitation on the exercise of state powers when state actions affect interstate commerce. In situations where Congress has exercised its power to regulate interstate commerce, the commerce clause combines with the supremacy clause to prevent inconsistent state action. In other situations, where Congress has neither expressly nor implicitly assumed a position on a particular matter affecting interstate commerce, the "dormant" commerce clause—that is, the unexercised commerce power and the policy favoring the free flow of commerce—limits state action that impinges on interstate commerce.[48]

In the latter situation, state action is usually challenged either as discriminating against interstate commerce or as imposing an undue burden on interstate commerce. Cases involving discrimination against interstate commerce ordinarily involve a state law that favors local businesses or products at the expense of out of state businesses or products.[49] None of the economic disincentives discussed in this book are designed to provide the type of economic protectionism that could render them vulnerable to being struck down as discriminating against interstate commerce.

The other ground for invalidating non-preempted state regulation of commerce—that it unduly burdens interstate commerce—looms as a tougher obstacle to economic disincentives. In determining whether a state action does unduly burden interstate commerce, courts weigh the local benefits of the law against the burdens it imposes. In *Pike* v. *Bruce Church, Inc.*,[50] the Supreme Court formulated the balancing test that courts should apply:

> Although the criteria for determining the validity of state statutes affecting interstate commerce have been variously stated, the general rule that emerges can be phrased as follows: Where the statute regulates even-

handedly to effectuate a legitimate local public interest, and its effects on interstate commerce are only incidental, it will be upheld unless the burden imposed on such commerce is clearly excessive in relation to the putative local benefits. Huron Portland Cement Co. v. City of Detroit, 362 U.S. 440, 443. . . . If a legitimate local purpose is found, then the question becomes one of degree. And the extent of the burden that will be tolerated will of course depend on the nature of the local interest involved, and on whether it could be promoted as well with a lesser impact on interstate activities.[51]

In this balancing process, greater weight should be accorded the state interest if the state regulation involves health or safety concerns as opposed to economic matters.[52] In some instances, however, the Court may not require the weighing process if the case involves non-comparable benefits and burdens.[53]

In the only litigated commerce clause challenge to an economic disincentive, the Oregon Court of Appeals, which did not apply the *Pike* balancing test, held that the state's bottle bill did not violate the commerce clause.[54] Of the economic disincentives analyzed in this book, the congestion pricing concept (discussed in Chapter 4) appears most likely to be challenged on commerce clause grounds. Assuming that a balancing test is applied, the local benefits of reduced congestion, air pollution, noise, and energy consumption will be weighed against the economic burden of paying a fee for traveling to congested places, along congested routes, at congested times. The burden seems reasonable, and not "clearly excessive," in relation to the benefits.

The Fourteenth Amendment. The fourteenth amendment provides that a state shall not "deprive any person of life, liberty, or property, without due process of law; nor deny to any person within its jurisdiction the equal protection of the laws."[55] The due process and equal protection clauses require that state actions accord with fundamental notions of reasonableness and fairness. These guarantees do not, however, empower courts to sit in judgment of the wisdom of either the policies determined by state legislators to be in the public interest or the legislation designed to implement those policies.[56]

The due process concept has a number of facets. Procedural due process grants individuals the right to notice and hearing when government actions threaten to deprive them of life, liberty, or property. Due process also prohibits the government from taking private property without paying just compensation.[57] As far as economic disincentives are concerned, due process "demands only that the law shall not be unreasonable, arbitrary or capricious, and that the means selected shall have a real and substantial relation to the object sought to be obtained."[58] This standard, however, does not present a great barrier to disincentives, because the Supreme Court has not invalidated any state economic regulation legislation on substantive due process grounds since the Depression. Today, the requirement of substantive due process will be considered satisfied if the means are rationally related to any legitimate state interest. Economic disincentives are certainly rationally related to the legitimate state interest in energy conservation.

Equal protection challenges to economic disincentives are probable if a disincentive strategy bears more heavily on one class of persons than another. The U.S. Supreme Court, however, has consistently allowed states and localities to draw reasonable distinctions among members of a class as long as the distinctions do not impair fundamental rights (e.g., voting or interstate travel) and are not based upon suspect classifications (e.g., race). In upholding a local police power enactment, the Court stated that:

> This Court consistently defers to legislative determinations as to the desirability of particular statutory discriminations. . . . Unless a classification trammels fundamental personal rights or is drawn upon inherently suspect distinctions such as race, religion, or alienage, our decisions presume the constitutionality of the statutory discriminations and require only that the classification challenged be rationally related to a legitimate state interest. States are accorded wide latitude in the regulation of their local economies under their police powers, and rational distinctions may be made with substantially less than mathematical exactitude. Legislatures may implement their program step by step . . . adopting regulations that only partially ameliorate a perceived evil and deferring complete elimination of the evil to future regulations.[59]

As for classifications in state tax legislation, the Court allows the states an even greater measure of discretion:

> The broad discretion as to classification possessed by a legislature in the field of taxation has long been recognized. . . . [T]he passage of time has only served to underscore the wisdom of that recognition. . . . Traditionally classification has been a device for fitting tax programs to local needs and usages. . . . It has, because of this, been pointed out that in taxation, even more than in other fields, legislatures possess the greatest freedom in classification. Since the members of the legislature necessarily enjoy a familiarity with local conditions which this Court cannot have, the presumption of constitutionality can be overcome only by the most explicit demonstration that a classification is a hostile and oppressive discrimination against particular persons and classes.[60]

More succinctly, the Court has noted that the equal protection clause "imposes no iron rule of equality, prohibiting the flexibility and variety that are appropriate to reasonable schemes of state taxation."[61]

Whether enacted under the police power or the tax power, none of the economic disincentives discussed in this book are susceptible to being successfully challenged on equal protection grounds. None of the strategies involve classifications that are based upon inherently suspect distinctions such as race, religion, or alienage. Moreover, none of the classifications impinge upon any of the fundamental personal rights.[62] All of the classifications, such as those suggested for the congestion pricing and parking tax proposals, can be easily defended as being rationally related to the legitimate state interest in energy conservation.

The foregoing analysis of the legal aspects of economic disincentives suggests several tentative conclusions. First, the police and taxing powers of state and local governments are broad and flexible enough to permit them to enact effective economic disincentives. Second, the limitations imposed on state and local government action by the supremacy clause, the commerce clause, and the fourteenth amendment of the U.S. Constitution do not present overwhelming barriers to the economic disincentives discussed in this book. Third, the key to bringing economic disincentives within the scope of state and local powers and to avoiding violations of the federal constitutional limitations is careful legislative drafting. Finally, before disincentive advocates can be sure that a particular disincentive strategy is free of legal defects, they must research state or local law to ascertain whether any additional limitations beyond those examined in this section present obstacles.

DESIGN ASPECTS OF ECONOMIC DISINCENTIVES

After advocates of economic disincentives have determined that a state or locality has the power to enact disincentives and before they begin the legislative drafting process, they must answer a series of design questions. This section provides a general overview of the choices designers face in resolving these questions. More detailed analyses of these issues are included in the discussions of the disincentives examined in Chapter 4. The major design questions for economic disincentives are:

- On whom should the disincentive be imposed?
- What basis should be used to calculate the amount of the disincentive?

- Should the disincentive be imposed gradually or all at once?
- How long should the disincentive be kept in effect?
- What should be done with the revenue that the disincentive generates?

The starting point in designing any economic disincentive is a determination of who will be subject to the disincentive. In deciding whether to impose a disincentive across the board on everyone or to limit it to a particular group of persons, designers should consider the disincentive's purpose—that is, whether the disincentive is intended to raise the cost of energy to encourage conservation generally or whether it has a more specific intent, such as discouraging a particular energy-consuming activity for which there are less energy-intensive alternatives. In the former case, imposing the disincentives on all energy users seems fairest and most consistent with the objective of encouraging everyone to reduce their energy use.

In the later case, however, applying a disincentive across the board may not be fair. Moreover, subjecting everyone to the tax may not achieve significantly greater energy savings than would limiting it to a specific group. Consider, for example, a parking tax whose purpose is to encourage people to use public transportation or to form carpools instead of continuing to drive by themselves. Public transportation service is best and opportunities for carpools are greatest during rush hours. At other times, people may have no reasonable alternative to using their cars. Limiting the tax to persons who park their cars during the morning rush hours would be fairer than imposing it on everyone, including those persons who have no choice but to drive.

The question of who should be subject to paying a particular disincentive raises the issue of exemptions. Broad exemptions could significantly dilute any disincentive's conservation impact, but narrow exemptions may be justified out of consideration for fairness or administrative simplicity. For instance, fairness may dictate that handicapped persons who need specially equipped cars to transport themselves be exempted from a parking tax. Fairness does not require, however, an exemption for persons for whom a disincentive would cause only economic hardship. Other measures could be devised that would alleviate the burden of paying the tax or fee without disturbing the disincentive's conservation effect. For example, instead of exempting the poor from paying congestion fees, subsidized public transportation could be provided so that no one, including poor persons, would have to rely on a car for traveling to and within the congested area. Administrative simplicity may also dictate the creation of exemptions. An automobile efficiency tax–rebate program, for instance, would probably be most effective if it applied to used cars as well as to new cars. The difficulty of determining the tax or rebate for models that do not have fuel efficiency ratings, however, suggests that used cars should be exempted from the program.[63] Because they have the potential for becoming huge loopholes, exemptions should be regarded warily, used sparingly, and drawn narrowly.

Having determined who will pay, designers must decide the crucial issue of how much the disincentive will be. In selecting a basis for calculating the amount of the disincentive, the key factor will again be the purpose of the disincentive under consideration.

If the disincentive is intended to raise the cost of energy to encourage conservation, several bases exist for computing the amount of the disincentive. One basis, discussed briefly in Chapter 1, is the replacement cost of energy. Replacement costing of energy is favored by many experts as the best means for eliminating the deterrent to conservation that arises from the underpricing of energy.[64] Energy is considered underpriced because the prices consumers pay are lower than the costs of replacing the energy consumed. For example, under federal price controls, the price of domestic crude oil ranges from $5.51 to $12.21 per barrel and averages $9.05 per barrel.[65] The replacement source, foreign crude oil, cost over $12.70 per barrel in 1978.[66] As a result of price controls, the market sends consumers a signal that oil costs roughly $10.50 per barrel (the approximate composite cost per barrel of domestic and foreign oil). Naturally, consumers make fewer conservation efforts and investments when they believe oil costs $10.50 per barrel than they would if they received price signals that reflected the replacement cost of oil.

The federal government has greater opportunities than the states for correcting this problem because federal price controls on domestic crude oil and natural gas are largely responsible for the underpricing of energy. Several of the taxes proposed in the National Energy Plan,[67] such as the crude oil equalization tax, would have provided a partial solution to the underpricing problem. Along with many of the other effective parts of the Plan, however, the crude oil equalization tax was not passed by Congress as part of the National Energy Act.

Although the states cannot change federal price controls, they do have several opportunities for correcting the underpricing of energy. For example, electric utility rates could be revised to reflect the marginal or long run incremental, rather than average, costs of providing electricity.[68] Of the economic disincentives discussed in this book, the gasoline tax is the one that could be most easily calculated and imposed on a replacement cost basis.

The cost of energy could also be raised (and disincentives calculated) on the basis of the externalities associated with energy production and use. Externalities are costs arising from an activity that are borne by society at large rather than by those who benefit from the activity. Among the externalities related to energy production are health damage (e.g., black lung disease) and environmental degradation (e.g., oil spills). The use of energy generates automobile congestion and pollution (air and noise), air pollution resulting from power generation by utilities and industries, and pollution-related health and property injury. These adverse consequences of energy production and use impose tremendous costs on society, but the price of energy includes only a percentage of these costs. For states and localities, problems such as the difficulty of determining the value of some externalities limit the practicality of using this approach to calculate disincentives. Moreover, the primary objective of the disincentives discussed in this book is to encourage energy conservation. Computing these disincentives on the basis of pollution externalities may produce effective pollution control, but poor energy conservation, measures.

If, instead of trying to encourage conservation generally by raising the cost of energy, a disincentive is intended to achieve a particular purpose (e.g., to influence people to buy more efficient cars), designers will face the task of trying to determine what amount or rate will induce the desired result. Unfortunately, the lack of actual experience with disincentives will make the process of determining the proper amount a guessing game. In most instances, however, designers will have some information that will reduce the degree of guessing involved. For some disincentives,[69] studies by the government or by private research organizations have estimated the impact of imposing a disincentive in varying amounts. In addition to these types of studies, public opinion polls, surveys of consumer buying habits and preferences, and related experience[70] can provide guidance concerning how the public may react to a particular disincentive. Of course, none of these sources of information will provide definitive answers to the question of whether imposing a disincentive of a certain amount will, in a particular state or locality, produce the desired result. Ultimately, selecting the amount of a disincentive that will be effective in achieving its purpose will involve some degree of trial and error.

A final basis for calculating the amount of a disincentive is the amount of money an individual or business saves by not undertaking mandated conservation activities. A state could, for instance, promulgate thermal efficiency standards for existing buildings. To encourage building owners to retrofit their buildings promptly, the state could enact a disincentive that would be imposed when owners delayed compliance with the standards. The amount of the disincentive would be based on the savings realized by the owner as a result of his noncompliance with the standard. Knowing that when they are caught they will lose any savings they may have gained from noncompliance, many building owners, lacking the incentive for delay, will probably comply with the standards within the prescribed time. This idea represents an application of the concept underlying the economic civil penalties provision of Connecticut's pollution control law. The appendix to this chapter briefly describes the Connecticut system.

After determining the amount of the disincentive, designers need to select a time frame for its introduction. The choice involves whether to impose the disincentive in its full amount at one time or whether to phase it in gradually. If the disincentive is relatively small (e.g., the 5 cent deposit imposed by bottle bills), imposing the full amount all at once will cause no hardship. If, however, the disincentive is more substantial (e.g., a $1 per gallon gasoline tax or a $2 per day parking tax), designers should consider imposing the disincentive gradually to give people time to make the adjustments that the disincentive is designed to encourage (e.g., driving less and relying upon public transportation more). The risk of phasing in the disincentive is that the public will react as they do to inflation—with complaints, but without action. Too many people may respond to disincentives that are phased in by paying them rather than by changing their behavior, so that the disincentive ultimately has little effect on energy use. Because of this danger, disincentives should, in most cases, be imposed in their full amount after allowing the public an adequate period of time to change their energy consumption habits.[71]

Not only must designers decide how to introduce a disincentive, but they must also decide how long the disincentive will remain in effect. A disincentive could be imposed for an indefinite period, for only the time necessary to achieve certain goals, or for a limited term. In determining the life span of a disincentive, designers should keep in mind that disincentives are intended to achieve *and maintain* goals (e.g., pricing energy at its replacement cost or reducing automobile use by *x* percent or encouraging recycling) and that some disincentives will be relied on for their revenue.

Enacting disincentives for an indefinite period runs the risk that they will remain in effect after the need for them has passed (although the need for revenue may never pass). Tying disincentives to the accomplishment of certain objectives raises the danger that the goals will not be maintained after the disincentives expire. Arbitrarily limiting the existence of a disincentive to a specific term ignores the possibility that a disincentive may still be needed even after the term has ended. Perhaps a requirement for periodic legislative review to determine whether the disincentive has been effective and whether the need for it remains is the best way to ensure that disincentives do not expire before their time and do not outlive their usefulness.

The final major design question concerns the disposition of the revenue that almost every disincentive will generate. Chapter 2 recommends that, to enhance the political feasibility of disincentives, advocates of these measures should propose spending the money in ways that alleviate the negative aspects of disincentives. Thus, revenues can be spent to cushion the impact of disincentives on those who can least afford to pay them (the poor) and to finance less energy-intensive alternatives to the activities on which the disincentives are imposed. Chapter 4 discusses specific examples of these suggestions.

NOTES TO CHAPTER 3

1. *See generally* F.R. Anderson et al., *Environmental Improvement through Economic Incentives* (Baltimore: The Johns Hopkins University Press, 1977); *Pollution Taxes, Effluent Charges, and Other Alternatives for Pollution Control*, a report prepared by the Congressional Research Service of the Library of Congress for the U.S. Senate Committee on Environment and Public Works, Serial No. 95–5, (Washington, D.C.: U.S. Government Printing Office, May 1977); A.V. Kneese and C.L. Schultze, *Pollution, Prices, and Public Policy* (Washington, D.C.: The Brookings Institution, 1975); W.A. Irwin and R.A. Liroff, *Economic Disincentives for Pollution Control: Legal, Political and Administrative Dimensions*, EPA–600/5–74–026, prepared for the U.S. Environmental Protection Agency by the Environmental Law Institute (Washington, D.C.: Environmental Protection Agency, July 1974).

2. Courts have not hesitated to strike down poorly drafted environmental charges. *See, e.g.,* Society of the Plastics Industry, Inc. v. City of New York, 68 Misc. 2d 366, 326 N.Y.S. 2d 78 (Sup. Ct. 1971) (the New York City recycling incentive tax was invalidated partly because the city ordinance failed to comply with the provisions of the state enabling act); McClain v. Board of Supervisors, No. 4156 (Cir. Ct. of Loudon County, Va., 1976) (the Loudon County bottle bill was invalidated as unconstitutionally vague because it did not define the key concept of minimum deposit).

On the other hand, numerous courts have upheld carefully designed environmental charges and economic disincentives. *See, e.g.,* American Can Co. v. Oregon Liquor Control Comm., 15 Or. App. 618, 517 P. 2d 691 (1973) (bottle bill);

Bowie Inn, Inc. v. City of Bowie, 274 Md. 230, 335 A.2d 679 (1975) (bottle bill); Andrews v. Lathrop, 132 Vt. 256, 315 A.2d 860 (1974) (land gains tax); City of Pittsburgh v. Alco Parking Corp., 417 U.S. 369 (1974) (parking tax).

3. Munn v. Illinois, 94 U.S. 113, 145 (1877).

4. Noble State Bank v. Haskell, 219 U.S. 104, 111 (1911) (citation omitted).

5. Block v. Hirsh, 256 U.S. 135, 155 (1921).

6. Lawton v. Steele, 152 U.S. 133, 137 (1894).

7. Goldblatt v. Town of Hempstead, 369 U.S. 590, 595 (1962), citing Sproles v. Binford, 286 U.S. 374, 388 (1932).

8. American Can Co. v. Oregon Liquor Control Comm., 15 Or. App. 618, 517 P.2d 691, 698 (1973).

9. 517 P.2d at 700.

10. 4 T.M. Cooley, *The Law of Taxation*, 4th ed. (Chicago: Callaghan & Co., 1924), § 1784 at 3513.

11. *Id.*

12. *Id.* at 3514.

13. 4 Cooley, *supra* note 10, § 1809 at 3555. *See, e.g.,* Jordan v. Village of Menomonee Falls, 28 Wis.2d 608, 137 N.W.2d 442 (1965) (upholding a $200 per lot fee that subdividers could pay in lieu of dedicating land in the subdivision for school, park or recreational purposes).

14. *See, e.g.,* Skidmore v. City of Elizabethtown, 291 S.W.2d 3 (Ky. 1956) (rejecting a challenge to parking meter rates that produced excess revenue: "[I]t is clear that the matter of parking, both on-street and off-street, is all part of the main traffic regulation problem, and therefore there is nothing improper in utilizing excess revenues from the parking meters to meet the costs of the overall traffic regulation police problem, or in fixing the parking meter fees at an amount that will produce such excess revenues." 291 S.W.2d at 5).

15. 4 Cooley, *supra* note 10, § 1809 at 3555. *See, e.g.,* Lyons v. City of Minneapolis, 241 Minn. 439, 63 N.W.2d 585 (1954); Sager v. City of Silvis, 402 Ill. 262, 83 N.E.2d 683 (1949).

16. Allied Stores of Ohio, Inc. v. Bowers, 358 U.S. 522, 526 (1959).

17. Sonzinsky v. United States, 300 U.S. 506, 513 (1937).

18. United States v. Sanchez, 340 U.S. 42, 44 (1950) (citations omitted).

19. Magnano Co. v. Hamilton, 292 U.S. 40, 44 (1934) (citation omitted).

20. City of Pittsburgh v. Alco Parking Corp., 417 U.S. 369 (1974).

21. 417 U.S. at 375 (citations omitted).

22. Andrews v. Lathrop, 132 Vt. 256, 315 A.2d 860 (1974). The Land Gains Tax, Vt. Stat. Ann. tit. 32, §§ 10001–10010, is a tax imposed on the profit derived from the sale or exchange of land held by the transferor for less than six years. The tax rate is in direct proportion to the percentage of gain and in inverse proportion to the holding period. *See generally* R.L. Baker, "Controlling Land Uses and Prices By Using Special Gain Taxation to Intervene in the Land Market: The Vermont Experiment," 4 *Envt'l Aff.* 427 (1975); Note, "State Taxation—Use of Taxing Power to Achieve Environmental Goals: Vermont Taxes Gains Realized From the Sale or Exchange of Land Held Less than Six Years—Vt. Stat. Ann. tit. 32, §§ 10001–10 (1973)," 49 *Wash. L. Rev.* 1159 (1974).

23. Andrews v. Lathrop, 132 Vt. 256, 315 A.2d at 863.

24. *See generally* D. J. McCarthy, Jr., *Local Government Law* (St. Paul, Minn.: West Publishing Co., 1975), pp. 14–23; 1 C.J. Antieau, *Municipal Corporation Law*, ch. V (New York: Matthew Bender, 1975); D. Francis, "Energy Conservation: Guidelines for Municipal Programs," 8 *Envt'l L.* 131, 132–34 (1977).

25. Merriam v. Moody's Ex'rs., 25 Iowa 163, 170 (1868) (Dillon, C.J.). *See also* 1 Antieau, *supra* note 24, § 5.01 at 5–3.

26. *See generally* McCarthy, *supra* note 24 at 23–42; 1 Antieau, *supra* note 24, §§ 5.17–5.25; 5.35–5.43; Francis, *supra* note 24 at 144–45, 150–52. Local police power enactments may be challenged on the grounds that they are preempted by state law. State constitutional limitations on municipal tax powers include requirements that taxes be (1) uniform or equal, (2) in proportion to the value of the property, (3) levied for a local purpose, and (4) levied for a public purpose. These limitations may not apply to all types of municipal taxes, and they are subject to varying interpretations from state to state.

27. McCarthy, *supra* note 24 at 105. For a list illustrating the broad range of activities that municipalities have dealt with under the police power, *id.* at 110–11.

28. *Id.* at 105–06. When the challenge alleges improper exercise, courts apply the same test of reasonableness that they use to judge state police power enactments; Francis, *supra* note 24 at 134–46.

29. McCarthy, *supra* note 24 at 227; Francis, *supra* note 24 at 149.

30. Florida Lime & Avocado Growers, Inc. v. Paul, 373 U.S. 132, 142–43 (1963).

31. California v. Zook, 336 U.S. 725, 728–29 (1949); Cloverleaf Butter Co. v. Patterson, 315 U.S. 148, 156 (1942); Hines v. Davidowitz, 312 U.S. 52, 67 (1941); Northern States Power Co. v. Minnesota, 447 F.2d 1143, 1154 (8th Cir. 1971), *aff'd mem.*, 405 U.S. 1035 (1972).

32. Campbell v. Hussey, 368 U.S. 297, 302 (1961); Hood & Sons, Inc. v. Du Mond, 336 U.S. 525, 543 (1949).

33. Florida Lime Avocado Growers, Inc. v. Paul, 373 U.S. 132, 147–50 (1963); Campbell v. Hussey, 368 U.S. 297, 301–02 (1961); Northern States Power Co. v. Minnesota, 447 F.2d 1143, 1149 (8th Cir. 1971), *aff'd mem.*, 405 U.S. 1035 (1972).

34. Florida Lime & Avocado Growers, Inc. v. Paul, 373 U.S. 132, 143–44 (1963); San Diego Building Trades Council v. Garmon, 359 U.S. 236, 241–44 (1959); Northern States Power Co. v. Minnesota, 447 F.2d 1143, 1153–54 (8th Cir. 1971), *aff'd mem.*, 405 U.S. 1053 (1972).

35. Hines v. Davidowitz, 312 U.S. 52, 67 (1941).

36. Pennsylvania v. Nelson, 350 U.S. 497, 502–04 (1956); Rice v. Santa Fe Elevator Corp., 331 U.S. 218, 230 (1947); Bethlehem Steel Co. v. New York State Labor Relations Bd., 330 U.S. 767, 774 (1947).

37. Florida Lime & Avocado Growers, Inc. v. Paul, 373 U.S. 132, 145 (1963); Kelly v. Washington, 302 U.S. 1, 10 (1937); Chemical Specialties Mfr's Ass'n v. Clark, 482 F.2d 325 (5th Cir. 1973) ("For where Congress has chosen to 'occupy' a field, but has not undertaken to regulate every aspect of that area,

the states have the implied reservation of power to fill out the scheme." 482
F.2d at 327 citing Florida Lime & Avocado Growers v. Paul, *supra.*)

38. *See* Merrill, Lynch, Pierce, Fenner & Smith, Inc. v. Ware, 414 U.S. 117,
127 (1973); and Colorado Anti-Discrimination Comm'n v. Continental Air
Lines, Inc., 372 U.S. 714, 722 (1963).

39. *See* Askew v. American Waterways Operators, Inc., 411 U.S. 325, 331
(1973).

40. P. L. 94—163, § 509, codified at 15 U.S.C. 2009.

41. P. L. 94—163, § 327 as amended by the National Energy Conservation
Policy Act of 1978, P. L. 95—619, § 424, codified at 42 U.S.C. § 6297.

42. P. L. 94—163, §§ 362—366 (state energy conservation plans provisions);
P. L. 94—385, §§ 431—432 (supplemental state energy conservation plans)
codified at 42 U.S.C. §§ 6322—6327.

43. National Energy Conservation Policy Act of 1978, P. L. 95—619, §§
621—622 authorizing $100 million in federal assistance for fiscal year 1979.

44. To be eligible for assistance, state plans must include (1) mandatory
lighting efficiency standards for public buildings; (2) programs to promote the
availability and use of carpools, vanpools, and public transportation; (3) mandatory standards and policies relating to energy efficiency to govern the procurement practices of the state; (4) mandatory thermal efficiency standards and
insulation requirements for new and renovated buildings; and (5) a right turn on
red traffic law or regulation. 42 U.S.C. § 6322(c).

45. 42 U.S.C. § 6322(d).

46. U.S. Const. Art I, § 8, cl. 3.

47. *See* Heart of Atlanta Motel v. United States, 379 U.S. 241 (1964); Katzenbach v. McClung, 379 U.S. 294 (1964); Wickard v. Filburn, 317 U.S. 111
(1942).

48. *See* G. Gunther, *Cases and Materials on Constitutional Law*, 9th ed.
(Mineola, New York: Foundation Press, 1975), p. 277.

49. Pike v. Bruce Church, Inc., 379 U.S. 137 (1970) (invalidating a requirement that Arizona cantaloupes be packed in Arizona); Dean Milk Co. v. City of
Madison, 340 U.S. 349 (1951) (invalidating preferential inspection laws); Bald-
win v. G.A.F. Seelig, Inc., 294 U.S. 511 (1935) (invalidating milk quota preferences for in-state milk).

50. Pike v. Bruce Church, Inc., 397 U.S. 137 (1970).

51. *Id.* at 142 (parallel citations omitted). This test was recently applied by
the Court in Great Atlantic and Pacific Tea Co. v. Cottrell, 424 U.S. 366, 371—
72 (1976).

52. Dixie Dairy Co. v. City of Chicago, 538 F.2d 1303, 1308 (7th Cir.
1976).

53. *See, e.g.,* American Can Co. v. Oregon Liquor Control Comm., 15 Or.
App. 618, 517 P.2d 691, 697—98 (1973), relying upon analysis in Firemen v.
Chicago, R.I. and P.R. Co., 393 U.S. 129, 136, 138—39 (1968).

54. American Can Co. v. Oregon Liquor Control Comm., 15 Or. App. 618,
519 P.2d 691 (1973). Although the court did not apply a balancing test, some
academic commentators have concluded that if the test were applied the local
benefits would be found to outweigh the burden. *See, e.g.,* Note, "State Envi-

ronmental Protection and the Commerce Clause," 87 *Harv. L. Rev.* 1762 (1974). *Contra* amicus curiae brief in *American Can* filed by Professors Philip Kurland, Alexander Bickel, and Gerald Gunther.

55. U.S. Const. amend XIV, § 1.

56. *See* City of New Orleans v. Dukes, 427 U.S. 297, 303 (1976); Nebbia v. New York, 291 U.S. 502, 537 (1934).

57. For the likely disposition of claims that economic disincentives enacted as taxes violate the due process protection against taking property without compensation, *see* Magnano Co. v. Hamilton, 292 U.S. 40, 44 (1934), and City of Pittsburgh v. Alco Parking Corp., 417 U.S. 369, 373−74 (1974), discussed in Chapter 4.

58. Nebbia v. New York, 291 U.S. 502, 525 (1934).

59. City of New Orleans v. Dukes, 427 U.S. 297, 303 (1976), citing Lehnhausen v. Lake Shore Auto Parts Co., 410 U.S. 356 (1973); Katzenbach v. Morgan, 384 U.S. 641 (1966); Williamson v. Lee Optical Co., 348 U.S. 483 (1955) ("Evils in the same field may be of different dimensions and proportions, requiring different remedies. Or so the legislature may think. Or the reform may take place one step at a time; addressing itself to the phase of the problem that seems most acute to the legislative mind. The legislature may select one phase of the field and apply a remedy there, neglecting the others." 348 U.S. at 488−89).

60. Madden v. Kentucky, 309 U.S. 83, 87−88 (1940), quoted in San Antonio School District v. Rodriguez, 411 U.S. 1, 40−41 (1973).

61. Allied Stores of Ohio, Inc. v. Bowers, 358 U.S. 522, 526 (1959).

62. The right to travel is the only fundamental right that is even conceivably affected by any of the economic disincentives discussed in this book. Despite the motoring public's belief to the contrary, the right to travel does not encompass a right to travel by automobile. *See* Note, "Legal Aspects of Banning Automobiles from Municipal Business Districts," 7 *Column. J. L. Soc. Prob.* 412, 431 (1971). As long as congestion pricing plans and parking taxes do not unduly restrict the mobility of commuters, courts should not subject the classifications made in the plans and taxes to strict scrutiny.

63. Excluding all used cars from an efficiency tax rebate program would appear, at first glance, to create an unnecessarily broad exemption that would encourage people to buy inefficient older cars. Chapter 4, however, recommends that annual registration fees be revised to serve as disincentives for owning heavy (and in most cases, inefficient) vehicles. This strategy, therefore, would at least partially offset the effect of the suggested exemption.

64. Raising energy prices to reflect the true replacement cost of energy was one of the underlying principles of President Carter's National Energy Plan. Executive Office of the President, Energy Policy and Planning, *The National Energy Plan* (Washington: U.S. Government Printing Office, 1977), pp. 29−30.

65. Energy Information Administration, U.S. Department of Energy, *Monthly Energy Review* (Washington, D.C., November 1978), p. 59. In August 1978, $5.51 was the wellhead price of lower tier oil, $12.21 the wellhead price of upper tier oil, and $9.05 the actual domestic average.

66. In December 1978 the OPEC decided to boost oil prices in 1979 by 14.5 percent. As a result of this decision, the price of imported oil will rise to $14.50 per barrel by the end of 1979.

67. *The National Energy Plan, supra* note 64.

68. *See* F.J. Wells, *Utility Pricing and Planning—An Economic Analysis* (Cambridge, Massachusetts: Ballinger Publishing Co., 1979).

69. *See, e.g.*, higher gasoline taxes, congestion pricing, and parking taxes discussed in Chapter 4.

70. *E.g.*, designers of automobile efficiency tax rebate programs should study the automaker's programs using rebates to encourage sales of slow-selling small cars during the 1974—1975 recession and price increases to discourage sales of big cars. In December 1978, Ford, Chrysler, and General Motors raised the prices of their least fuel-efficient models because the heavy demand for those models was jeopardizing their chances of reaching the sales-weighted fleet fuel economy average mandated by the Energy Policy and Conservation Act (15 U.S.C. § 2002).

71. One obvious exception to this recommendation involves disincentives designed to serve as standby measures. For example, President Carter proposed a standby gasoline tax as part of his National Energy Plan. *See The National Energy Plan, supra* note 64 at 38—39. Under this proposal, federal gasoline taxes would have increased 5 cents after each year in which motorists exceeded national gasoline consumption targets. If a particular year's consumption fell within the targeted level, gasoline taxes would have been reduced 5 cents the next year. Congress never showed any interest in this proposal.

Appendix to Chapter 3

CONNECTICUT ECONOMIC CIVIL PENALTIES

Since 1973 the Connecticut Department of Environmental Protection has had the authority to impose economic civil penalties on pollution sources that violate either the state's pollution standards or the terms of pollution permits or abatement orders.[1] The economic penalties are based on the actual savings realized by a source as a result of its noncompliance with the standard, permit, or order. The savings are determined through calculations involving two objective variables—the cost of compliance and the length of time compliance has been delayed.

The primary purpose of the Connecticut penalties is to make the regulatory system more effective by encouraging voluntary compliance. The penalties are designed to accomplish this objective by making compliance a justifiable investment. If businesses invest in compliance, they have something to show for the expenditure (e.g., pollution abatement equipment). If they pay the penalties, they save nothing and have nothing to show for the expenditure.

The penalties are also designed to force the most intransigent polluters into compliance. One study found that between June 1972 and mid-1974, twenty-one Connecticut sources of air pollution (less than 10 percent of the sources receiving abatement orders during the period) accounted for 75 percent of the sixty to seventy aggregate years of delay resulting from failures to meet the orders' compliance deadlines.[2] The civil assessments are designed to eliminate any savings gained from these delays. The civil penalties are also intended to protect complying sources from being placed at a competitive disadvantage relative to their noncomplying competitors.[3]

This appendix briefly describes the statutory provisions authorizing imposition of the penalties and the enforcement process outlined in the regulations governing their application in the state's air pollution control program.[4]

The enabling statute for the penalties, the Enforcement Act of 1973,[5] empowers the commissioner of the Department of Environmental Protection to adopt schedules of penalties for four categories of violations and places ceilings on the penalties for each category. The law also enumerates factors that the commissioner must consider in devising the schedules and in determining the penalty in a particular case. For example, in setting the civil penalty in a particular case, the commissioner is directed to consider the following factors.[6] :

1. The amount of assessment necessary to ensure immediate and continued compliance;

2. The character and degree of impact of the violation on the natural resources of the state, especially any rare or unique natural phenomena;

3. The conduct of the person incurring the civil penalty in taking all feasible steps or procedures necessary or appropriate to comply or to correct the violation;

4. Any prior violations by such person of statutes, regulations, orders, or permits administered, adopted, or issued by the commissioner;

5. The economic and financial conditions of such person;

6. The character and degree of injury to, or interference with, public health, safety, or welfare that is caused or threatened to be caused by such violation;

7. The character and degree of injury to, or interference with reasonable use of, property that is caused or threatened to be caused by such violation.

The statute also details the procedures that the commissioner must follow in levying a penalty, grants polluters the right to seek judicial review of final administrative orders, and exempts persons who are complying with the terms of orders and permits from being subject to civil penalties.

The administrative process under the air regulations begins when the commissioner, after detecting violations of air emissions standards, orders, or permits, sends a source a warning letter. By acting promptly to correct the violations specified in the warning letter, the recipient of the letter, unless he is a repeat violator or already had notice of the specific violation and potential penalty, may avoid assessment of the penalty. If the warning letter is ignored, the discharger will receive a formal notice identifying the violation and the maximum potential penalty and informing him of his right to a hearing. Following the hearing, the commissioner will issue a final order; or if the polluter does not request a hearing, the notice automatically becomes a final order twenty days after its issuance.

Following the entry of a final order, the commissioner can calculate and impose the civil penalty. The regulations contain a schedule of maximum monthly penalties based upon the equipment and operating expenditures necessary to achieve compliance. Each individual polluter's penalty is calculated by:

1. Estimating the cost of the equipment and operating changes necessary to achieve compliance (by consulting cost manuals developed by the Department of Environmental Protection and/or using accepted engineering estimating techniques);

2. Using a formula to convert these costs into the savings the source would realize for each month's delay in compliance;[7] and

3. Multiplying these monthly savings by the number of months the source has delayed compliance.

This computation determines an individual source's maximum assessment for a particular violation. The commissioner, however, has the discretion to reduce the penalty upon consideration of factors such as the financial condition of the polluter.[8] In addition to this potential mitigation of the penalty, the source is entitled to a refund with interest, if after achieving compliance, the polluter demonstrates that the amount of the assessment exceeded the actual cost of compliance.[9]

Imposing the civil penalty is the most severe sanction available to the commissioner under the Enforcement Act. One advantage of the Connecticut approach, its flexibility, is reflected in the other less severe economic enforcement tools to which the commissioner may resort in his discretion.[10] For example, instead of levying an assessment, the commissioner can allow a polluting company to sign an escrow agreement in the amount, for instance, of the value to the company of a year's delay in compliance.[11] If the existence of a signed, but unsecured, escrow agreement is not sufficient to induce the company to begin compliance, the commissioner can then require the posting of a percentage of the total value of the agreement, known as the reserve requirement, as security. Progress toward compliance would free some of the security; further delay would produce an increased reserve requirement and, eventually, forfeiture of the security. Connecticut has found the escrow agreement tool particularly effective in reducing delays in compliance. Following the signing of escrow agreements, two companies which had been behind in their air pollution compliance schedules by 66 percent and 133 percent reduced the delay to 2.5 percent and 0 percent, respectively.[12]

Not only do civil penalties eliminate the profit of delay and protect law-abiding companies, but they also reduce the administrative costs of enforcement. With the incentive for delay removed, voluntary compliance increases, and the number of lengthy enforcement investigations and proceedings decreases. Because unappealed administrative penalties are collectible as judicial judgments, a cumbersome collection process is avoided. It is possible, therefore, that the use of penalties could enable a pollution control agency to reduce the size of its enforcement and litigation staffs or to direct its energy to problems that have not been adequately addressed in the past.

The example discussed in the text suggests one area in which economic civil penalties could be used to encourage energy conservation measures. As disincentives to delay, economic civil penalties should be considered whenever conservation programs establish standards and a deadline for meeting them.

NOTES TO APPENDIX

1. *See generally Economic Law Enforcement*, 6 vols., EPA—901/9—76—003a, prepared for the U.S. Environmental Protection Agency Region I, by the Connecticut Enforcement Project, (Washington, D.C., September 1975). In addition to being authorized for violations of standards, permits, and orders, civil penalties are also authorized for violations of a variety of procedural requirements, such as the requirements concerning the submission of progress reports.

2. D.W. Tundermann, "Economic Enforcement Tools for Pollution Control: The Connecticut Plan" (Paper presented at the Environmental Study Conference [U.S. Congress] briefing, "Tax and Fee Approaches to Pollution Control," May 26, 1976), and reprinted in *Pollution Taxes, Effluent Charges, and Other Alternatives for Pollution Control*, a report prepared by the Congressional Research Service of the Library of Congress for the U.S. Senate Committee on Environment and Public Works, Serial No. 95—5 (Washington, D.C.: Government Printing Office, May 1977), pp. 835—57.

3. *Id.*

4. Regulations of Connecticut State Agencies [hereinafter Regulations] §§ 22a—6b—602 and 22a—6b—603. In March 1977, regulations authorizing the

imposition of penalties for violations of the state's water permit requirements became effective.

5. Public Act 73–665 codified at Conn. Gen. Stat. § 22a–6b and § 22a–6(7).

6. Conn. Gen. Stat. § 22a–6b(c).

7. *See Economic Law Enforcement*, vol. 1, *supra* note 1 at 56–65.

8. Regulations §§ 22a–6b–602(g)(1) and 22a–6b–603(g)(1).

9. Regulations §§ 22a–6b–602(g)(2)(ii) and 22a–6b–603(g)(2)(ii).

10. Public Act 73–665 § 3(7) codified at Conn. Gen. Stat. § 22a–6(7).

11. *See* Tundermann, *supra* note 2; and *Economic Law Enforcement*, vol. 1, *supra* note 1, at 14–16.

12. *Economic Law Enforcement*, vol. 1, *supra* note 1, at 15.

Chapter 4

Economic Disincentives
and the Automobile

INTRODUCTION

During the era of inexpensive energy that lasted from the end of World War II to the fall of 1973, Americans fell in love with the automobile. Cheap gasoline fueled the romance, and the automobile's charms cemented the bond. Cars served as universal status symbols. For many years, keeping up with the Joneses required buying newer and bigger cars. Automobiles also provided flexibility and convenience. Drivers could step out of their homes or offices at any time and soon be on the road to virtually any destination. In addition to mobility, cars gave commuters an isolation booth away from family, co-workers, and other commuters. Nothing threatened Americans' love affair with their automobiles until the oil embargo.

To cope with the embargo, the government examined various opportunities for reducing petroleum consumption. Passenger and freight transportation, accounting for one-half of the nation's petroleum consumption, became a prime target of conservation efforts. The automobile, burning one-third of the oil used in America in 1973, became the bull's eye on the target.

The government took aim at the automobile with measures designed to yield immediate gasoline savings. Highway speed limits, for example, were lowered from sixty-five or seventy miles per hour (mph) to fifty-five. In addition to trying to get people to drive more slowly, the government tried to get people to drive less. Public service advertisements urged motorists to join carpools or to use mass transit, drivers could only buy gasoline on odd- or even-numbered days, and service stations closed on Sundays. Combined with sharply higher gasoline prices, these policies succeeded in substantially reducing gasoline consumption.

When the embargo ended, lines at service stations disappeared; gasoline soon became plentiful, although at almost twice the pre-embargo price; and Americans gradually became reconciled with their automobiles. Some government officials knew that despite the virtual resumption of business as usual, a serious energy problem remained, and they began to formulate long range energy programs.

In developing a program to reduce automotive energy consumption, the government can choose from among policies that reflect two basic approaches: (1) strategies to reduce the number of automobile vehicle miles traveled (VMT) and (2) strategies to increase automobile efficiency. The Energy Policy and Conservation Act (EPCA), which Congress passed in late 1975, utilizes both approaches.[1] EPCA limits imposition of some VMT reduction measures, such as rationing and commuter parking restrictions, by the federal government to times of "severe energy supply interruptions."[2] It leaves other automobile use restrictions to the states to enact (e.g., transportation controls as part of the state energy conservation program).[3] In contrast to this cautious approach to measures that would directly affect individuals, Congress acted decisively in establishing a program to increase automobile efficiency. Forcing the automakers to improve fuel economy has undeniable political appeal—a more efficient car requires no individual sacrifice (i.e., less driving) and promises the same or greater personal mobility for fewer gallons of gasoline.

EPCA amends the Motor Vehicle Information and Cost Savings Act to add provisions establishing mandatory automobile fuel economy standards.[4] For the 1978 model year the standard is 18 miles per gallon (mpg), for 1979 it is 19 mpg, for 1980 it is 20 mpg, and for 1985 it is 27.5 mpg. As required by EPCA, the secretary of transportation has established standards for 1981 (22 mpg), 1982 (24 mpg), 1983 (26 mpg), and 1984 (27 mpg).[5] A Federal Energy Administration (FEA) study[6] estimates that without these fuel efficiency standards the industrywide sales-weighted fuel economy of new car fleets would have increased from 16 mpg for the 1975 model year to 24.2 mpg for the 1985 model year. With the standards, the increase will be from 16 to 28.9 mpg. The study also estimates that in 1985, 400,000 barrels of oil will be saved daily as a result of the mandatory fuel economy program.

According to the U.S. Environmental Protection Agency (EPA), eleven of thirteen foreign and domestic automakers met the 1978 standard with their 1977 models and seven of the eleven met the 1980 standard.[7] Beginning with the 1978 model year, however, any manufacturer whose production-weighted fleet fuel economy for a model year does not equal or surpass the standard for that year is subject to a civil penalty.[8] The civil penalty is calculated by multiplying the total number of cars that the automaker produced in the model year by $5 for each tenth of a mile per gallon that the actual fleet average falls below the EPCA standard. Provisions of the National Energy Act (NEA) authorize the secretary of transportation to promulgate regulations increasing the penalty up to $10 for model years after 1980.[9]

Unfortunately, EPCA did not couple its potentially tough standards and noncompliance penalties for manufacturers with comparable demands on new car buyers. Rather than directly encouraging consumers to buy more energy-efficient cars, EPCA simply seeks to ensure, through its labeling requirement, that prospective buyers are provided with fuel economy information, which they may ignore when making their decisions.[10] The labeling provision requires each car manufactured in model years after 1976 to bear a sticker indicating:

1. the car's fuel economy;

2. the estimated annual cost for fuel to operate the automobile; and

3. the range of fuel economy for comparable automobiles (manufactured by the same and other manufacturers).[11]

By the end of 1976, automobile sales figures, showing intermediate and full size models increasing their share of the market at the expense of more fuel-efficient imported and domestic models,[12] and record summer gasoline sales convinced many people that informational labeling alone would not encourage the public to buy more efficient cars or save very much gasoline. Among those convinced was President Carter who proposed, as part of his National Energy Plan, a tax on inefficient, and a rebate on efficient, new cars and a standby gasoline tax.[13] The rebate and gasoline tax ideas never received serious consideration in Congress, and the gas guzzler tax survived as a largely symbolic measure.[14] The failure of Congress to pass a strong federal program to reduce automotive energy consumption, however, preserves significant opportunities for state and local initiatives.

States have several options for complementing the federal efforts—
the fuel economy standards and gas guzzler tax—to improve the fuel
efficiency of the nation's automobile fleet. First, states *may* be able
to enact their own automobile efficiency taxes and rebates. (See Appendix 4—1 for a discussion of whether the federal gas guzzler tax
preempts similar state programs.) Second, whether or not state efficiency taxes and rebates are preempted, states can revise their annual
automobile registration fees so that fees for efficient cars are lower
than those for inefficient cars. Efficiency improvement policies such
as these will reduce gasoline consumption to a degree. Not until they
are combined with programs that reduce the number of miles people
drive, however, will the full potential for automobile-related energy
conservation be within reach.

VMT reduction measures include a wide range of strategies, from
programs to promote carpools, vanpools, and other forms of ridesharing to rationing and parking supply restrictions to taxes on fuel
and on automobile use. Except for a restriction on gasoline or diesel
fuel rationing,[15] Congress has left the imposition of these various
transportation controls up to the states. This chapter analyzes three
economic disincentives designed to reduce VMT—higher gasoline
taxes, congestion pricing schemes, and parking taxes. Not only could
these measures help states to achieve EPCA's energy conservation
goals;[16] they can also assist states and localities in dealing with local
problems such as the deleterious effects of urban motor vehicle congestion and the lack of funds for developing and supporting transit
and paratransit services.[17]

AUTOMOBILE FUEL EFFICIENCY
TAXES AND REBATES

In the years following the embargo, Congress and numerous state
legislatures considered proposals to impose energy efficiency taxes to
discourage the production and purchase of energy-wasteful automobiles.[18] None of the states ever enacted automobile efficiency taxes,
but in October 1978, Congress passed a gas guzzler tax as part of the
National Energy Act.[19]

The federal gas guzzler tax will be initially imposed on 1980
model year automobiles. Automobiles whose fuel economy is greater
than 15 mpg will not be subject to any tax. For 1980 cars that get
less than 15 mpg, the tax will range from $200 to $550. Each year
the "zero tax" fuel economy level will creep higher. By the 1986
model year, the tax rates will range from $500 (for cars with ratings

between 21.5 and 22.5 mpg) to $3,850 (for cars with ratings of less than 12.5 mpg).

These taxes will probably affect only a few cars each model year. In the 1980 model year, for example, EPCA's fuel economy standard for the manufacturers is 20 mpg. To avoid the penalties for failure to meet the standard, manufacturers will probably not produce many models that have ratings below 15 mpg. In addition to focusing on only the most egregious gas guzzlers, the taxes offer no encouragement to people to buy the relatively more efficient models. The shortcomings of the federal gas guzzler tax suggest the need for state action in this area. The legal question this action would raise—preemption—is discussed in Appendix 4–1. The conclusion of the analysis of this issue is that the federal tax does not bar similar state taxes.

State programs, however, should consist of more than taxes to discourage the purchase of inefficient cars. Some studies have suggested that a tax alone may, because of its depressing effect on new car sales, actually impede the transition to a more efficient stock of cars.[20] Taking a cue from the manufacturers who have successfully resorted to rebate programs to move slow-selling small cars, states could complement the taxes with rebates to encourage the purchase of efficient models.

A bill introduced in the Colorado House of Representatives in 1976 to establish a tax-rebate program[21] provides a useful point of reference for analyzing the tax-rebate proposal. (See Appendix 4–2 for the text of the bill.) The Colorado bill requires a determination of the "energy efficiency" of each new car registered in the state to be made at the time of its initial registration. The "energy efficiency" of a particular model is the model's mpg rating for combined city and highway driving as calculated in tests by the EPA.[22] Under the tax-rebate schedule in the bill, owners of new cars with energy efficiencies below 20 mpg would pay an "ownership" tax that increases from $20 to $320. Purchasers of new cars with energy efficiencies above 21 mpg would receive a "payment" of between $20 and $300. All taxes would be deposited in an energy conservation reserve fund, and all rebates would be paid from the tax revenues placed in the fund.

Two aspects of the Colorado bill merit special attention. First, the tax is designed for maximum impact and visibility. By requiring purchasers to pay the tax at the time of vehicle registration rather than at the time of sale, the legislature has avoided the problem of the tax being "hidden" as another item on the sales contract and financed as part of the purchase price. Furthermore, this arrangement neatly frustrates buyers who might travel out of state to buy a new car in order to avoid a tax levied at the time of sale. Under the proposed Colorado arrangement, purchasers of inefficient cars cannot escape having to shell out an extra $200, for example, to register their new cars initially.

Second, the Colorado tax-rebate program, like virtually all other proposed automobile efficiency taxes, applies only to purchasers of new cars. Purchasers of used ("previously owned") cars are not subject to the tax or eligible for the rebate. Perhaps recognizing that inefficient used cars waste gasoline just as inefficient new cars do and that failure to tax inefficient used cars could stimulate the demand for them, partially defeating the goal of the tax-rebate program, legislation proposed in Arizona would tax used cars at one-half the rate applicable if the car were new.[23] A tax-rebate program for used cars, however, may be difficult to administer, particularly in the case of older models that do not have an EPA mileage rating. Calculating the tax or rebate for these models on the basis of vehicle weight, a fairly reliable surrogate for fuel economy, might solve this problem. In any event, because rebates for efficient new cars and relatively high registration fees for inefficient used cars may offset the effect on demand for used cars caused by not taxing them, a legislature should carefully investigate the need for extending the tax-rebate program to used cars before doing so.

Key provisions of any tax-rebate law are those dealing with the schedule of taxes and rebates, the disclosure (e.g., by labeling) of the tax-rebate schedule, and the disposition of the tax revenue. The ways in which these issues have been handled in the various proposed bills suggest a number of points that legislators should consider in drafting these provisions of tax-rebate legislation.

Schedule of Taxes and Rebates

In developing a tax-rebate scale, a legislature must decide two basic issues. First, what point on the fuel economy spectrum should be chosen as the "base" fuel economy rating (i.e., the point at which the tax ends and the rebate begins). Second, how large a tax or rebate is necessary to accomplish the program's objectives.

The Colorado bill selects 20 to 21 mpg as its base fuel economy rating. Other examples of auto efficiency tax legislation, none of which include a rebate provision, drop the tax at points ranging from 15 to 30 mpg. One Maryland bill proposes a $20 tax for each tenth of an mpg that the car's fuel mileage rating falls below the applicable EPCA fuel economy standard (i.e., 19 mpg in 1979 and 20 mpg in

1980).[24] States should, in any case, not set the base rating below the applicable EPCA standard, and in view of the relatively modest EPCA standards in early years, states should consider choosing a base rating higher than the applicable standard (e.g., 24 mpg instead of 20 mpg for 1980). Naturally, the higher the base rating, the stronger will be the pressure on consumers to buy the more efficient models; and the greater the demand for these models, the easier it will be for the manufacturers to meet the production-weighted fleet fuel economy standards imposed on them by EPCA.

In addition to relating the tax-rebate schedule to the federal government's efforts to improve fuel economy, a state should develop a schedule that reflects the manufacturers' achievements in improving fuel economy. Consider the problem of determining at which fuel economy rating to apply the maximum tax. Under the Colorado proposal only cars that have a fuel efficiency rating of less than 5 mpg would be liable for the maximum tax of $320. According to EPA's estimates of the fuel economy of 1979 models, however, very few new cars have ratings of less than 15 mpg.[25] Unless the tax-rebate program keeps pace with the actual improvements in fuel economy, it will lose some of its potential impact.

One way of helping the tax-rebate schedule to maintain its effectiveness is to include an acceleration or escalator clause in tax-rebate legislation. An escalator clause would provide for periodic increases in the base fuel economy rating. The following sample acceleration clause is based on a provision in a proposed Maryland bill:[26]

> Beginning January 1, 19___ [the first or second year following enactment], each fuel economy rating in the tax-rebate schedule shall be increased every year [or every two years] by two miles per gallon.

After the first application of such a clause to the tax-rebate schedule in the Colorado legislation, the base fuel economy rating would be 22 to 23 (instead of 20 to 21) mpg, the maximum tax would fall on vehicles with fuel efficiency ratings of less than 7 (instead of 5) mpg, and to qualify for the maximum rebate, a car would need a rating of 37 (instead of 35) mpg.

An escalator clause, however, has purposes other than forestalling erosion in the tax-rebate program's effectiveness. First, if, as in the Colorado bill, the program is intended to be self-sufficient (i.e., rebates being paid exclusively from efficiency tax revenues), the revenue fund may eventually be depleted as the average fuel economy of the new car fleets improves. An escalator clause should prevent complete depletion. Second, an acceleration clause will provide an additional incentive for manufacturers to develop and market more efficient cars.

Most of the proposed auto efficiency tax bills would impose the same tax on every car that has the same fuel economy rating. An alternative, which is much less equitable, would apply a variable percentage tax (i.e., the tax rate increases as fuel economy decreases) to the sales price of the automobile. The problem with this proposal is illustrated by the following example. Suppose one person buys a $7,000 car that has a fuel economy rating of 25 mpg while another person buys a $5,000 car that gets 20 mpg. Under one percentage tax proposal, the tax rate on cars getting 25 mpg would be 2 percent, and the rate on 20 mpg cars would be 3 percent.[27] Applying these rates to the sales prices in this example would result in the purchaser of the more efficient, but higher priced car paying a higher tax than the buyer of the less efficient auto. This result seriously undermines the justification for a fuel efficiency tax program. Even if this problem could be remedied (e.g., by changing the tax rates), a program in which the amount of the tax or rebate depends exclusively on the car's fuel economy rating should be less confusing to consumers and easier to administer than a scheme in which the tax varies according to the sales price of the car as well.

The second series of issues that legislatures must confront in devising a tax-rebate schedule involves the amount of the taxes and rebates. Essentially, legislators must decide how onerous taxes must be to discourage consumers from buying inefficient autos and how generous the rebates should be to influence people to purchase efficient cars. An examination of the various fuel efficiency tax proposals demonstrates that no consensus on these questions has emerged. The taxes in an Arizona bill, for example, ranged from $100 to $1,000, but one Connecticut bill would have imposed taxes that increased from $10 to only $90.[28] The gas guzzler taxes proposed in the National Energy Plan ranged from $52 to $2,488; the taxes passed by Congress vary from $200–550 for the 1980 model year to $500–3,850 for the 1986 model year.[29] Colorado and Connecticut bills, which provide for rebates as well as taxes, included rebates that went up to $300; the National Energy Plan proposed rebates of up to $500.[30] The wide divergence demonstrated by these various bills suggests that no one really knows what tax and rebate rates will be effective. Perhaps surveys of consumer buying preferences and attitudes could provide a rational base from which to proceed in devising tax and rebate rates.

Another point to be considered in deciding tax and rebate levels is the relationship of the two schedules. The scales in the Colorado bill are virtual mirror images. Taxes range from $20 to $320 and rebates from $20 to $300. The manufacturers' experience with rebates suggests that relatively modest rebates, such as those proposed in Colorado, may be effective in stimulating sales of efficient cars. Given the obduracy of consumer demand for inefficient models, however, a legislature should consider making the taxes much higher than the rebates. For example, the various tax rates in the Colorado bill could be doubled or tripled while the rebate amounts remain unchanged. Additionally or alternatively, the differentials between the steps in the tax scale could increase progressively instead of remaining the same. Thus, in the Colorado bill, the differentials, which are a constant $20, could be changed to increase progressively by a factor of $10 for each step up the tax scale (e.g., $20, $30, $40, and so forth). Revising the proposed Colorado tax scale in either of these ways will produce a stiff schedule of taxes to discourage the purchase of inefficient vehicles. In certain instances, no tax, no matter how steep, will deter a person from buying a particular model—a luxury car, for example.[31] For many consumers, however, the stiff taxes or attractive rebates could significantly influence their purchasing decisions.

Disclosure of Tax-Rebate Information

In order for the tax-rebate program to exert its influence on consumers' buying preferences, consumers must be made aware of the potential tax or rebate consequences their decisions will have. Mandatory labeling provisions and other types of disclosure requirements will ensure that this vital information is made readily available to prospective new car buyers.

Only a few of the proposed state auto efficiency tax bills have included labeling provisions. The following sample provision is based on clauses contained in proposed Maryland legislation:[32]

Every automobile dealer shall display on each new automobile on or immediately adjacent to every label required to be affixed under section 3 of the Federal Automobile Information Disclosure Act (15 U.S.C. § 1232) the following information:

(1) the fuel economy rating determined to be applicable for the automobile, which appears on the label required by 15 U.S.C. § 2006 (a)(1)(A);

(2) the tax or rebate applicable for the automobile; and

(3) the entire schedule of taxes and rebates applicable for the model year.

A label presenting this information makes the consumer aware of not only the consequences of buying the particular model he is considering but also the range of tax or rebate options that are available. States should carefully draft any labeling requirements concerning disclosure of fuel economy ratings to make the state mandate identical to EPCA's labeling requirement.[33] A conflict between the state and federal labeling requirements could result in the state provision being challenged by dealers and struck down by a court on preemption grounds.

The Colorado legislation takes a different approach to the information problem. It includes a provision requiring dealers to disclose the applicable tax or payment in any document (e.g., contract or estimate) or any advertisement that states the automobile's purchase price or acquisition cost. The Colorado information provisions do not apply to contracts, advertisements "or other disclosure involved or appearing in items of interstate commerce." Requiring disclosure of the applicable tax or rebate in advertisements and contracts would further enhance consumers' awareness of the financial impact of their decisions and would effectively complement labeling requirements.

Disposition of Revenue

The Colorado bill creates an energy conservation fund into which all ownership taxes are deposited and from which all rebates are paid. In the event that the fund becomes temporarily depleted, rebates are guaranteed, but payment is delayed until more tax revenues are deposited in the fund. After rebates are paid to eligible owners and counties are reimbursed for administration expenses, any surplus revenues are transferred into the state's general fund.

By restricting rebate payments to efficiency tax revenues, a legislature in effect forces the program to become self-supporting and prevents any drain on general revenues. As discussed above, including an escalator clause in the law is one way to avoid jeopardizing the goal of self-sufficiency. Making the taxes higher than the rebates should also help to make the program self-supporting.

If the tax-rebate program does generate some surplus revenue, the legislature has several options for spending it. While the Colorado proposal would transfer all of it to the general fund, a proposed Arizona luxury tax on motor vehicles would channel 50 percent of the tax revenues to solar energy research! Because one of the impediments to reducing this nation's reliance on the automobile is the absence of alternative modes of transportation, states could earmark all, or a portion, of the surplus revenues to finance transit and paratransit services. In some states, however, this proposal may face hurdles in constitutional or statutory provisions that dedicate motor vehicle excise taxes and registration fees exclusively to highway purposes. Each state will have to determine whether revenues from a fuel efficiency tax would be controlled by its dedication provision.

Administration of a fuel efficiency tax-rebate program by a state's departments of motor vehicles and revenue should not present any major difficulties or costs. The administrative process for car registration is well established, and adding the step required by this bill (i.e., receipt of the tax or payment of the rebate) should not unduly complicate the process. The administrative costs of the tax-rebate program can be covered by the tax revenues.

Enforcement of the program should not be a problem. An owner's failure to pay the tax simply precludes registration of the car. To ensure that dealers comply with the labeling and/or other disclosure provisions, the tax-rebate statute could include a section similar to the following, which is based on a provision in a proposed Maryland bill: [34]

> Penalty:
>
> (1) Any automobile dealer who willfully fails to display or disclose the information required by section _____ [the labeling and/or disclosure section], or displays false information, shall be fined not more than $1,000 for each failure or false information display.
>
> (2) Any person who removes, modifies, or mutilates any information affixed [under the labeling provision] to an automobile before title or possession to the automobile is transferred to a final purchaser shall be fined not more than $1,000 for each removal, modification, or mutilation.

Apparently, no study has been conducted to determine the energy savings from a state tax-rebate program. [35] Common sense, however, suggests that the savings will be relatively modest in the short run because of the infrequent opportunities any program will have for making an impact. Most people buy a new car only once every five to ten years. Even then, a person could ignore the rebate lure and tax burden and select a new gas guzzler. Between trade-ins, gasoline prices provide the only existing incentive for owning more fuel-efficient cars. Recent experience indicates, however, that gasoline prices at current levels (65 to 75 cents per gallon) are a rather weak incentive. States, therefore, should consider revising their annual automobile registration fees to complement a tax-rebate program.

AUTOMOBILE REGISTRATION FEES

To fill in the gaps left by a tax-rebate program, states can adjust their schedules of annual registration fees for passenger automobiles so that the fee for a particular category of cars reflects the fuel economy generally achieved by that class of cars. Even without a tax-rebate program, registration fees based on energy efficiency criteria would serve as an effective yearly reminder that it pays to own an efficient car.

Ideally from an energy conservation perspective, the registration fee for a particular car should relate directly to its fuel economy. In actuality, registration fees in most states have little relationship to fuel economy. Twenty-two states charge a flat fee to register any car; twenty-three states impose fees that vary according to a vehicle's empty, shipping, or gross weight; two states use a car's horsepower as the basis for the fee; and three states use a car's age and/or value to determine its registration fee.[36] Because vehicle weight is a significant determinant of a vehicle's fuel economy, states could transform their schedules of automobile registration fees into energy conservation tools by enacting a schedule of fees based on vehicle weight.

A weight-based schedule of fees should have two features. First, it should have weight classes that correspond to the common car categories (i.e., subcompacts, compacts, intermediates, and full size). Although some cars in a particular category will achieve better gasoline mileage than others, a schedule that reflected these differences more precisely would be more difficult and costly to administer than the one suggested here. With the right scale of fees, a schedule limited to these four weight classes will provide an incentive for people to switch from larger cars to smaller ones. The second feature is a scale of fees that provides a progressive disincentive to the ownership of larger cars. This feature can be accomplished by making the differences in the four fees increasingly greater. These design features are reflected in the 1977 schedule of auto registration fees for the District of Columbia:[37]

Subcompacts (2,800 pounds or less)	$50
Compacts (2,801–3,499 pounds)	$57
Intermediates (3,500–3,999 pounds)	$83
Full size (4,000 pounds or more)	$96

In addition to serving as an impetus to the acquisition of more efficient automobiles, the higher registration fees would raise more revenue.[38] Subject to the constitutional and statutory limitations (which could be changed in those states where they exist) on the diversion of registration fees to nonhighway purposes,[39] these increased revenues could be used to finance transit and paratransit service in order to provide an alternative to the automobile.

Higher automobile registration fees, whether enacted separately or as part of a program including efficiency taxes and rebates, will probably not have an immediate impact on gasoline consumption. In fact, given that the entire automobile fleet in this country turns over approximately once every ten years, any strategy designed to improve the average fleet fuel economy, even if it accelerates the replacement rate, will only have a significant effect on gasoline consumption in the long run. Furthermore, this type of strategy does nothing to encourage people to drive less—a significant means of saving gasoline. The remainder of this chapter examines three measures—higher gasoline taxes, congestion pricing, and parking taxes—that can cut gasoline consumption by reducing automobile VMT.

GASOLINE TAXES

Before the embargo, cheap gasoline (30 cents per gallon) fueled Americans' love affairs with their cars. During the 1960s and early 1970s gasoline consumption grew at an average annual rate of 3.5 percent. The embargo, however, which brought short supplies and higher prices, temporarily disrupted the affair. In 1974, gasoline consumption fell 3.8 percent from the record 1973 level.[40] After the embargo, the shortages disappeared, but prices remained high (55 to 60 cents per gallon). For a while, the higher prices restrained demand. In 1975, consumption increased only 2.5 percent over 1974.[41] Gradually, however, the public adjusted to the higher prices, and although prices crept upward, demand soared. In 1976, consumption leaped 6.2 percent, reaching a record level; in 1977 consumption grew 3.4 percent; and in 1978, according to preliminary estimates, consumption rose 4 percent.[42] Increased gasoline consumption, combined with other factors such as unusually harsh winter weather, have kept the United States too dependent on foreign oil. Although the government has little control over the vagaries of the weather, it does have the power and the tools to affect the gasoline consumption patterns of American drivers.

A system of gasoline rationing is indisputably the most effective strategy for immediately reducing gasoline consumption. A plan limiting each licensed driver to ten gallons of gasoline a week would reduce consumption practically overnight by nearly 40 percent.[43]

The mere mention of rationing, however, usually raises howls of protest and elicits horror stories detailing the administrative nightmares that past rationing efforts have produced. Even during the embargo, the public and their elected officials fiercely resisted suggestions of rationing. Perhaps only during a truly serious national emergency would the hardships imposed by rationing become acceptable.[44] In addition, one important consideration for states is that gasoline rationing is more appropriately a federal, rather than a state, strategy.

An alternative to rationing, which the states as well as the federal government can implement, is a substantial increase in the excise taxes levied on gasoline. Since 1959, the federal excise tax has been 4 cents per gallon. State excise taxes range from 5 to 11 cents per gallon.[45] Total federal and state taxes represent less than 20 percent of the cost of a gallon of gasoline. In contrast, gasoline taxes in other industrialized nations, including Belgium, West Germany, Italy, Norway, Sweden, Switzerland, and the United Kingdom, account for over 50 percent of the price of a gallon of gasoline.[46] Table 4-1 compares the total price of a gallon of gasoline with the per gallon taxes imposed in these and other nations. Europe's high gasoline prices, resulting primarily from high taxes, have been at least partly responsible for keeping per capita gasoline consumption lower in Europe than in the United States.

Table 4-1. Gasoline Prices and Taxes in Industrial Countries *(as of December 31, 1975)*.

Nation	Total Price per Gallon[a] (in cents)	Total Taxes per Gallon (in cents)
United States	59.1	12.1
Belgium	136.1	84.6
Canada	70.5	27.8
Germany	123.2	67.5
Italy	174.7	112.9
Japan	139.1	42.8
Norway	138.7	73.1
Sweden	129.6	71.3
Switzerland	143.2 (super)	82.7
United Kingdom	114.1	60.1

[a] All prices are for regular grade gasoline, except for Switzerland.

Source: Organisation for Economic Cooperation and Development, *Energy Conservation in the International Energy Agency—1976 Review*, (Paris, 1976), Tables 1 and 2.

Imposing substantially higher taxes on gasoline is perceived by the public to be an effective conservation strategy for this country. A Harris Survey conducted in early 1977[47] revealed that by 74 to 21 percent, the survey's respondents believed that a 50 cent per gallon increase in the price of gasoline would be an effective way to reduce gasoline consumption. Similarly, by a 70 to 25 percent margin, the poll's respondents felt that a 25 cent per gallon increase would be effective. Unfortunately, however, a strategy's effectiveness is often inversely related to its acceptability. The same survey found that only 15 percent of the respondents would favor increasing gasoline taxes to raise the price of gasoline to $1.00 per gallon. A resounding 77 percent opposed this conservation measure.

It is fairly obvious why such a substantial majority of the public opposes proposals to increase the price of gasoline. Most people feel that gasoline prices are already too high and that as a result, the oil companies are earning exorbitant profits. Sharply higher prices, even if they resulted from increased taxes, which would go to the government instead of the oil companies, raise fears of drastic lifestyle changes. These changes are, however, purely from a energy conservation perspective, quite desirable. Higher gasoline taxes will produce a reduction in VMT. Faced with significantly higher prices, most people will probably drive their own cars less and shift to other means of transportation, including transit, paratransit, bicycles, and feet, to make some trips they would formerly have made by car. Seeking to regain their lost, or maintain their threatened, mobility, many people will respond to higher gasoline taxes by purchasing more efficient cars. In this respect, higher gasoline taxes would complement the pressures exerted by the federal fuel economy standards and gas guzzler tax and by any state efficiency tax-rebate and registration fees programs.

The time may arrive, however, when despite public opposition, this country will have to accept either gasoline rationing or higher gasoline taxes. Besides the administrative and enforcement advantages of higher gasoline taxes, the tax approach also gives individuals the freedom to choose whether or not to change their lifestyles to conserve gasoline and save money. Although these factors may not make higher gasoline taxes more popular and acceptable to the public than rationing, policymakers should be aware of the key considerations involved in imposing higher gasoline taxes. The issues addressed in the remainder of this section are:

- The amount of tax increase, including whether to impose a substantial increase (e.g., 50 cents per gallon) all at once or gradually (e.g., 10 cents a year for five years);

- The type of tax—that is, whether to levy the tax as a fixed rate per gallon or as a percentage of the per gallon price; and

- The disposition of the revenues, including tax relief for the poor and financial support for transit and paratransit service.

The Amount of the Tax Increase

Two factors will enter into legislative considerations of raising gasoline taxes to promote energy conservation—the effectiveness of the proposed amount of increase in achieving gasoline savings and the public acceptability of the proposed levy. Not surprising, the more effective a proposed increase is likely to be, the less likely it is that the increase will be publicly acceptable. Realistically, therefore, public acceptability will be the controlling factor in decisions concerning the size of a gasoline tax hike and will operate to limit any increase. Despite the dim prospects for a substantial increase in the gasoline tax, legislators should be aware of the potential impact of different levels of increase.

Since 1973, economists and transportation planners have employed econometric modeling to study the demand for gasoline and to calculate the elasticity of demand for gasoline.[48] Elasticity of demand is expressed as a number that represents the percentage change in the demand for a product that will result from a 1 percent change in the price of the product. The estimates of elasticity of demand for gasoline vary widely. Projections of the short run (i.e., less than one year) elasticity of demand range from −0.06 to −0.83, meaning that a 50 percent increase in price could cut demand anywhere from 3 percent to 41 percent.[49] Most of the estimates of short run elasticity, however, fall between −0.06 and −0.18. Calculations of the long run elasticity of demand range from −0.07 to −0.92, meaning that a 50 percent increase in the price of gasoline could, after several years, reduce demand by 4 to 46 percent.[50] The variations in the estimates arise from the differences in the models employed to calculate the elasticities. Assessing the models to determine which most accurately reflects the factors determining elasticity of demand for gasoline is beyond the scope of this book and the expertise of its author.

In addition to expressing the effect of the price changes on the demand for gasoline in terms of the relatively esoteric concept of elasticity, these and other studies have estimated the impact on consumption of specific price or tax increases. A summary of some of these projections follows.

In a report that assessed the effect of various carpool incentives, including a gasoline price increase, for the Washington, D.C., area, Cambridge Systematics, Inc., estimated the effect on fuel consumption of doubling, tripling, and quadrupling the price of gasoline.[51] This study projected that doubling the price of gasoline would reduce gasoline consumption in Washington by 4.7 percent. Tripling the price would result in a 9 percent savings, and quadrupling the price would produce a 12.9 percent savings.

In a study of alternative policies for saving gasoline, the Rand Corporation calculated the effects of 15, 30, and 45 cent increases in the tax.[52] Assuming that the tax increase had been imposed nationwide in 1975, by 1980 the 15 cent hike would have reduced consumption 16.2 percent from the level of consumption that could be expected without the tax, the 30 cent increase would have cut consumption 33.8 percent, and the 45 cent increase would have slashed consumption 40.5 percent.[53] Considering another approach, a 50 percent increase in the price of gasoline, Rand estimated that the increase in its first year, assuming a preincrease price of 67 cents per gallon, would cut consumption annually by 34.6 percent and that in five years, assuming the price of 67 cents increased annually at the same pace as inflation, a 50 percent increase would produce yearly savings of 37.6 percent.[54]

A study prepared by Charles River Associates[55] assessed the impact on gasoline consumption of increases in the federal excise tax of 10, 25, and 50 cents a gallon. Assuming that higher taxes were imposed in 1975, the 10 cent levy would have reduced consumption that year by 3 percent, the 25 cent levy by 7.4 percent, and the 50 cent levy by 14.9 percent. Six years after enactment, the 10 cent tax would save 10.6 percent of the projected level of consumption, the 25 cent tax 26.6 percent, and the 50 cent tax 53.2 percent. By the twelfth year, the estimated base consumption would be cut 11.9 percent by the 10 cent tax, 29.6 percent by the 25 cent tax, and 59.6 percent by the 50 cent tax.

Despite the confusing lack of agreement among the estimates of the elasticity of demand for gasoline and among the projections of the effect on consumption of various price and tax increases, a few tentative generalizations can be made about the effectiveness of higher gasoline taxes as an energy conservation strategy. First, the demand for gasoline is relatively insensitive to price or tax hikes in the short run. A 10 percent increase in the price will probably only reduce consumption by 1 percent in the months immediately following the increase. Demand will probably continue virtually unabated because of the difficulties of quickly trading in the entire fleet of inefficient cars and changing driving patterns, particularly where no alternative means of transportation exist. According to one survey conducted to determine how people reacted to the gasoline shortages and price increases caused by the embargo and why they reacted as they did, the availability of gasoline influenced travel behavior more than the higher prices, which were found to have little effect on demand.[56] Second, according to most studies, over the long run, demand for gasoline will be much more responsive to price. Over the course of several years, people have greater opportunities to buy more efficient cars and to find alternative means of transportation. This generalization about the long term effect of higher prices, however, needs to be tested against the postembargo experience in this country. Despite an increase in the price of gasoline of over 75 percent between 1973 and 1976, gasoline consumption in 1976 surpassed the record set in 1973. In light of this statistic, one wonders whether people might not eventually adjust to a significant increase in gasoline taxes and frustrate the achievement of its objectives.

Given the inherent unpopularity of higher taxes of any kind, the results of the studies indicating that higher gasoline taxes may not produce the immediate energy savings one might otherwise expect and the long run effectiveness of relatively less painful strategies to improve automobile fuel economy, the question arises whether the gasoline tax should be raised at all. Several justifications for increasing gasoline taxes do exist. First, although a tax increase of even 50 cents per gallon may not result in overnight reductions in consumption, an increase of only 5 cents per gallon semiannually or annually will send a clear signal to the public that changes in the automobile's role in our society are now, and will become increasingly, necessary. Second, by increasing the gasoline tax in 5 cent increments over a period of years, the legislature would be gradually building an effective strategy for the time in the future when the need to conserve will be even more imperative than it is today. Third, even if higher gasoline taxes do not reduce consumption immediately, they will produce revenue, which could be used for a variety of purposes (discussed later), including financing alternative means of transportation. Having decided, for any of these reasons, to raise gasoline taxes, legislators must then consider the type of tax to use.

Type of Tax

Higher gasoline taxes can be levied in one of two ways—as a specific amount per gallon, like existing gasoline taxes, or as a percentage of the sales price, like general sales taxes. Each method has its advantages.

Enacting new gasoline taxes on a per gallon basis has the virtues of familiarity and simplicity. An increase in gasoline taxes may be more publicly acceptable if it is perceived as an increase in an existing tax, rather than as an entirely new tax. In addition, this approach will not require new collection and enforcement procedures and mechanisms and will not impose new tax-reporting burdens on gasoline retailers. Finally, a per gallon tax that increases the pump price of a gallon may be a better signal to consumers than a percentage tax tacked onto the final sales price.

Proponents of a percentage tax approach cite two advantages of this alternative. First, a percentage tax, unlike a per gallon levy, would allow revenues to increase, without legislative intervention, at roughly the rate of inflation. In periods of inflation, a per gallon tax decreases as a percentage of the total price of a gallon, and the revenue it yields, unless demand increases, fails to keep pace with the rise in the costs of the activities (i.e., highway construction and maintenance) it supports. In view of this advantage, at least one state, Maryland, has studied the possibility of replacing its per gallon tax with a percentage tax.

The second advantage of a percentage tax is that it reflects the differences between the various grades of gasoline. If regular, which requires less energy to refine, sells for 63 cents a gallon and premium for 70 cents, the price ratio is 9 to 10. Imposing an additional per gallon tax of 30 cents changes the ratio to 9.3 to 10, and people may be encouraged to shift to the more energy-intensive grade of gasoline. An equivalent percentage tax of 50 percent, on the other hand, does not change the ratio. If the possibility of massive shifts to the more energy-intensive grade materializes, the legislature could design per gallon taxes to discourage the shift. For instance, applying a 27 cent tax to regular and a 30 cent tax to premium would keep the ratio in the above example at 9 to 10.

If one assumes that an immediate, substantial increase in the tax on gasoline, regardless of whether it is imposed as a 30 cent per gallon increase or as a 50 percent increase in the price, is not politically feasible, the issue becomes whether small periodic increases in gasoline taxes should be imposed on a per gallon or percentage basis. By imposing 5 cent per gallon increases semiannually, the revenue yields will keep pace with all but the most rampant inflation; and the change in the price ratio between the grades of gasoline will be so gradual that any danger of an increase in demand for the more energy-intensive grade will be minimized.

Not only should a series of small periodic increases in the gasoline tax meet the objections of proponents of a percentage tax approach, it also has the administrative advantages resulting from the retention of the existing tax procedures. Even if after several of these increases people do not significantly change their consumption habits, having become insensitive to the higher prices, gasoline tax increases are nevertheless justified on revenue grounds.

Disposition of Revenue

Because even small increases in gasoline taxes will yield significant revenues, legislators should consider the options for spending this money. One obvious choice is to devote at least some of the funds to highway construction and maintenance, purposes for which gasoline tax revenues have traditionally been used. This allocation, which would undoubtedly be popular with members of the highway lobby, may actually be justified in view of the squeeze on highway trust funds arising from the decrease in revenues caused by reduced gasoline consumption in 1974 and 1975 and the increase in construction and maintenance costs caused by inflation.

A second option is to use some of the money to provide tax relief to low income individuals and families whose limited incomes will undoubtedly be strained by the tax. Some legislators and citizens will oppose increased gasoline taxes because of their perceptions of the income distributional impact of the higher levies. Many of these opponents will intuitively assume that higher gasoline taxes will be regressive—that is, that the taxes will take a larger percentage of income from poor households than from middle or upper income households. The vast majority of the studies addressing this issue, however, have found that higher gasoline taxes would probably not be regressive—that is, any increases in gasoline taxes would take roughly equal percentages of income from all households.[57]

Concluding that, in general, the tax increases would not be regressive does not mean that some form of tax relief is not appropriate. Despite the conclusions of the studies, some low income families will suffer severe hardship as a result of the higher taxes. Relief for the poor can take a variety of forms: a provision to exempt them from paying the higher taxes; a program of gasoline stamps (similar to the energy stamp proposals discussed in Chapter 5); income tax credits or deductions; or periodic cash rebates for taxes paid. Although each of these suggestions has potential problems, ranging from dangers of fraud and abuse to failure to provide timely relief, they demonstrate that solutions to the problem of the impact of higher gasoline taxes on the poor do exist.

One solution that would benefit not only the poor but all taxpayers would be to use the gasoline tax revenues to reduce sales and property taxes. As the next section discusses briefly, revenues from these taxes, which everyone pays, subsidize services that primarily benefit automobile drivers. This subsidy occurs because automobile-

related revenues from gasoline and other excise taxes are not sufficient to cover all automobile-generated costs. Using revenues from
higher gasoline taxes to reduce this subsidy would shift the responsibility for more of the automobile-generated costs from society as a
whole to those who produce the costs. Reducing the sales and property taxes would particularly benefit the poor, because these taxes,
unlike increased gasoline raxes, do have a regressive impact.

Another option for spending some of the new gasoline tax revenues is to finance transit and paratransit services in order to give people alternatives to the continued use of the automobile. Most states,
however, have constitutional ("good roads amendments") or statutory provisions dedicating their motor fuel taxes to highway purposes.[58] Amendment of these prohibitions will be necessary before
the revenues from increases in the gasoline tax can be diverted to, or
specifically earmarked for, public transportation purposes. In 1974,
Massachusetts voters approved a constitutional amendment permitting the expenditure of revenue from the highway fund for mass
transportation.[59] In 1976, Michigan passed legislation specifically
earmarking 0.05 cent of the state's 9 cent gasoline tax for public
transportation.[60] California has attempted to circumvent these prohibitions by extending the general sales tax to gasoline. Similarly,
Virginia enacted legislation in 1976 to allow its Washington, D.C.,
suburbs to impose the state sales tax on the sale of gasoline to
finance part of their share of the costs of the Washington area sub-
way system.[61] While strong resistance from the highway lobby to
amending or repealing these constitutional and statutory prohibitions
should be expected, the removal of these barriers to permit at least
some of the gasoline tax revenues to be used for badly needed transit
and paratransit projects may enhance the public acceptability of
higher gasoline taxes.

In conclusion, increasing gasoline taxes is examined as an alternative to gasoline rationing. It has both strengths and shortcomings as
an energy conservation strategy. While most studies have concluded
that raising gasoline taxes by even as much as 50 cents per gallon will
not dramatically reduce consumption in the short run, higher gasoline taxes will, according to many of these studies, produce significant energy savings in the long run. It is unlikely, however, that
many people will support or accept an immediate sizeable gasoline
tax increase, especially if it will not result in substantial conservation
for several years. A series of small increases in gasoline taxes (e.g.,
5 cents semiannually for five years), therefore, may be the only
politically feasible way to raise gasoline taxes. The periodic increases
would, on one hand, constitute a warning that the time for certain

lifestyle changes is drawing near and, on the other, gradually build into an effective long run strategy for that time. The higher revenues that come with each increase in the gasoline tax could, with the removal of legal barriers, be used to finance improvements in transit and paratransit services in order to lessen this nation's dependence on the automobile.

Despite the arguments for adopting this strategy for energy conservation reasons, opposition to higher gasoline taxes will probably overwhelm any support the proposal draws.[62] If the usual arguments raised against higher taxes (of any kind) for individuals are not sufficient to defeat the proposal, opponents can resort to a battery of other arguments. First, they can point to postembargo gasoline consumption patterns to argue that even in the long run, higher gasoline prices will not significantly and permanently reduce gasoline consumption. Second, they can argue that any state would be foolish to impose higher gasoline taxes unilaterally because such action would encourage people residing in border areas to cross state lines to buy gasoline in states with lower taxes. This exodus might not only ruin many gasoline retailers in the state with the higher taxes, but might also result in the state's collecting less gasoline tax revenue after raising taxes than it had previously. Finally, opponents can argue that higher gasoline taxes are inequitable in two ways: (1) the "rich" will be able to absorb the increased taxes while the poor and middle class will not; and (2) the higher levies will limit personal mobility by indiscriminately discouraging all automobile trips, even those for which no reasonable alternative means of transportation exist.

A more focused strategy, congestion pricing, concentrates the economic disincentive on driving at times (usually rush hours) and to places (usually the central business district) of motor vehicle congestion. Not only do commuting trips account for 40 percent of all gasoline consumed, but the opportunities for leaving the car at home and taking the bus or forming a carpool are greatest during rush hours and on trips to and from work. Unlike higher gasoline taxes, therefore, congestion pricing offers the promise of energy savings without a significant limitation on personal mobility.

CONGESTION PRICING

Among the factors responsible for the deplorable urban transportation situation is the underpricing of various public resources, including clear air and highway space, used or consumed by transportation activities. Economists regard these public goods as underpriced because the persons who use them, especially rush hour drivers of low occupancy automobiles, do not pay the full societal costs that using these resources generate. Most drivers are not aware of the costs to society of their actions, and many probably believe, erroneously, either that they have paid for the use of these resources or that these resources are free, to be used without charge. This misperception is at least partly responsible for the waste and overuse of these public goods.

A concrete example will bring this economic theory into focus. A study of the underpricing of these resources in the Los Angeles Air Basin[63] identified three categories of unperceived, or hidden, subsidies or costs: (1) hidden budgetary subsidies; (2) smog costs; and (3) congestion costs. The most significant hidden budgetary subsidies are the general tax revenues (collected from the entire public) that must be added to auto-generated revenues (i.e., fuel and registration taxes, sales taxes on auto parts, license fees, and fines) to cover total auto-related expenditures (i.e., highway maintenance, police and medical resources). The exemption of highways from property taxes and the lower property taxes enjoyed by parking facilities are other important hidden subsidies. Smog costs consist of the damage to health and property that results from auto air pollution. Congestion costs include both the costs of building additional highways to satisfy peak period (i.e., rush hour) demand and the costs of delay (i.e., the value of the time of travelers caught in traffic jams). According to this study, persons who drive during rush hours and periods of bad air pollution are responsible for smog and congestion costs of $1 billion a year in the Los Angeles area.[64]

For many years, economists and transportation planners have advocated congestion pricing policies as one solution to the problems created by the underpricing of highway space.[65] More recently, government officials have studied and promoted congestion pricing as a possible means for attacking automotive air pollution and energy consumption.[66] Generally, congestion pricing schemes involve imposing a fee on auto users for traveling at times and in areas of congestion, when and where the costs that these drivers generate are the highest. An "economically efficient" congestion fee would approximate the difference between the costs a congestion motorist imposes on society and his contribution toward these costs in registration fees, gasoline taxes, and the other costs of commuting. If an economically efficient fee were levied, drivers for whom the value of making the trip did not equal or exceed the true costs of making the trip would forego the trip, travel to a different destination or travel to the same destination by a different mode, at a less congested or lower priced time or along a less congested (and lower priced) route.

The end result of a congestion pricing policy—imposing an economically efficient fee that forced drivers to choose among these alternatives—would be the more efficient use of highway space.

Of course, congestion fees can be set to achieve objectives other than the assessment of automobile users for the costs they impose on society. For example, air pollution abatement and energy conservation goals can be served by a congestion pricing program. To accomplish these goals, the congestion fee might be based on achieving a certain reduction in automobile traffic. A fee set on this basis may be either lower or higher than the economically efficient fee under the circumstances.

Despite the promise of congestion pricing policies as effective tools for curbing automobile use, no American city has attempted to implement a sophisticated, comprehensive congestion pricing scheme.[67] Foreign cities, however, have been more receptive to congestion pricing plans. London considered, but never implemented, a program that would have required autos traveling in the city's core areas during peak times to display special licenses.[68] In June 1975, Singapore instituted a congestion pricing plan in its central business district (CBD) that doubled CBD parking fees and imposed a special license requirement for cars entering the CBD during peak hours (7:30—10:15 a.m.).[69]

In addition to raising parking fees and imposing the special license requirement, Singapore improved transit service, opened parking lots on the fringe of the CBD, and enacted stiff ownership taxes ($1,000) for cars with large engines (over 3,000 cubic centimeters), registration fees for new cars (10 percent of value), roadworthiness surcharges on all cars more than ten years old (110 percent of normal registration fees for cars ten to fifteen years old, 150 percent for cars more than fifteen years old), and import duties (45 percent of value). This campaign against the automobile has had dramatic results. Car sales and car ownership have declined. The number of private cars entering the CBD during the peak hours has dropped nearly 75 percent, from 43,000 to 11,000, with four-person carpools, which are exempt from the license requirement, accounting for over 40 percent of the 11,000. Daily bus ridership has increased 200,000 to 1.9 million in a city of 2.2 million.

American motorists, however, would never accept a program as tough as Singapore's, and most legislators would consider it political suicide to propose one. The congestion pricing aspects of Singapore's anticar campaign, however, are not totally unreasonable, and even if this country's political climate will not permit their implementation now, government officials and concerned citizens should at least be familiar with the most frequently discussed congestion pricing options.

Congestion pricing research has concentrated on four alternative schemes: (1) toll booth collection systems, (2) automatic vehicle identification (AVI) systems, (3) automobile-mounted meters, and (4) supplemental licenses. The characteristics, advantages, disadvantages, and costs of each of these alternatives will be described briefly.

Toll Booth Collection Systems

One way to implement a congestion pricing policy would involve erecting toll booths to collect fees at every point of entry into a congested area.[70] This alternative, although perhaps less objectionable to motorists than other forms of congestion pricing because of its similarity to existing bridge, tunnel, and highway tolls, may be impractical because of its lack of flexibility and its cost. Most urban areas would require hundreds of toll booths to cover every street leading into a congested area. To avoid this problem, toll booths could be built on only a few major roads with all the other streets closed off by barricades. Funneling traffic into a few main arteries, however, would probably produce long delays and greater energy consumption as vehicles slowed down and lined up to pay the tolls. Increasing the number of traffic lanes and collection booths at each toll plaza would reduce some of the congestion, but would consume more valuable urban land. Other problems may arise if the tolls are varied between peak and nonpeak times. If the price differential were large, considerable congestion could result as motorists waited for the shift to the lower fee. A series of gradual reductions in the peak fee, on the other hand, might produce significant confusion and administrative headaches. Another shortcoming of the toll booth alternative is that with fixed collection stations at the perimeter of the congested area, trips to and from points entirely within the area will completely escape paying their share of congestion costs.

In addition to all these drawbacks, the costs of a toll booth collection arrangement may be prohibitive. One study estimates that the construction costs of a single toll plaza with at least two approaching lanes would probably exceed $700,000.[71]

A further barrier to implementing a toll system for congestion pricing is a provision of federal law prohibiting tolls on federally aided highways.[72] This restriction, however, will not apply to the vast majority of streets in large cities, where only a small percentage of the road miles have been built with federal funds.[73] Tolls on bridges are subject to a confusing mass of federal, state, and local regulations.[74]

Automatic Vehicle Identification (AVI) Systems

All AVI systems operate in the same way: as a car travels past a particular point, that passage is registered and becomes the basis for assessing congestion prices.[75] Engineers have studied a variety of sophisticated AVI technologies, but a low frequency induction system, which has been tested on bus fleets in New York, New Jersey, and San Francisco,[76] is the most promising. This system involves equipping vehicles with identifying units. While the car is on the highway, its unit transmits signals to, or responds to signals from, a scanning or "interrogating" device, located at the roadside or embedded in the roadbed. Information identifying the vehicle, its location, and even the time of day could be transmitted to a central processing facility for preparation of a billing statement, similar to a long distance telephone bill.

Unlike a toll booth system, low frequency induction technology would not impede the flow of traffic, and the system's billing computer could easily calculate charges that varied according to the time of day or congestion conditions. The sophisticated billing potential, however, could constitute one of the most serious objections to AVI systems. Aside from being disgruntled by having to pay to use traditionally free highways, many motorists will oppose a system that records their movements as an invasion of their privacy. Strictly enforced procedures for safeguarding the information gained from the AVI system may alleviate some of these fears. Another feature of AVI systems, the requirement that all cars be equipped with the relatively expensive ($100) identifying units, creates a problem for out of town drivers and residents who rarely travel into the congested area. Finally, according to most estimates, a low frequency induction system involves relatively high initial equipment and operating costs.[77]

Automobile-Mounted Meters

Meters, similar to those used by taxicabs, are perhaps the most promising congestion pricing technology.[78] Vehicle-mounted meters would not only permit charges to be levied according to the time of day a car entered the congested zone, but would also allow charges to be calculated on the basis of the length of time a car was driven in the congested zone. Some of the meters would require activation by the driver when the car entered the congested zone, but others would operate automatically on signals from roadside or roadbed units. Some meters, like metered mail machines, would accumulate charges for periodic payment, and others would run on prepaid timed discs.

Extension lights on the car, indicating whether the meter was running, would aid in enforcement.

Although less expensive than the AVI systems, vehicle-mounted meters would present the same problems for out of town visitors and other drivers who seldom enter the congested area. The major obstacle, however, to the implementation of this congestion pricing technology is that the system components have not even been produced for experimental use.

Supplemental or Area Licenses

The congestion pricing alternative for which low cost technology is presently available is the supplemental or special area license.[79] To travel into and/or within a congested zone, drivers would have to purchase and display these licenses on their windshields. Although this system presents some administrative and enforcement difficulties, it avoids many of the problems associated with the other alternatives. Unlike the other three alternatives, a licensing program does not require special technology or large initial capital commitments. Unlike a toll collection system or an AVI system having pricing points only on the perimeter of the congested zone, a licensing approach does not allow cars traveling only within the congested area to avoid the charges. Unlike an AVI system, a licensing scheme does not threaten an individual's privacy. Finally, a licensing program does not create undue difficulty for out of town visitors or residents who drive into the congested area infrequently. Because a licensing scheme could be implemented at the present time without overwhelming problems, some of the important considerations relating to the implementation, administration, and enforcement of a licensing program will be briefly analyzed.

The first series of considerations involves the scope of the license program. Initially, program designers must decide which parts of the city to include in the license program. The Singapore area licensing scheme applies only to the CBD. In addition to the CBD, other congested areas, such as those around non-CBD employment centers, may also be included in the program. After determining the geographical boundaries of the program, planners must decide the time boundaries for the various areas—that is, what times of the day and days of the week will require a license. In Singapore, licenses are required on weekday mornings from 7:30–10:15 a.m. Another consideration related to the scope of the program is whether to include all vehicles in the license requirements or to exempt certain types of vehicles such as carpools, vanpools, and buses. The decisions concerning the program's scope could result in a program as simple as Singapore's or one much more complex, with license requirements and, thus, the amount of the congestion fee that vary from zone to zone, time to time, and vehicle to vehicle.

The next step in establishing an area license arrangement is to determine the amount of the license fee. One approach is to charge the estimated "economically efficient" fee; however, calculating the economically efficient level with precision may be difficult. Alternatively, program designers could, as Singapore has done, set a fee designed to achieve a certain reduction in traffic in congested areas during peak hours. Singapore charged $1.20 for a daily license initially and has raised the fee to $1.60. Technical studies of congestion pricing policies provide various methodologies for calculating congestion fees.[80]

Another series of considerations concerns the types of congestion licenses to issue. A municipality could issue annual, monthly, or daily licenses. Daily licenses, even if the motorist can purchase large quantities at a single time, would be more likely to heighten a driver's perceptions of the costs and energy consequences of his travel choices than monthly or annual licenses. Having to affix a different license every day would remind the driver of the purposes of the licensing program. The Urban Institute has studied three types of daily licenses.[81] Self-cancelling stickers, a futuristic alternative, would be activated when applied to the car's windshield. After a certain time (e.g., eight hours), when the license expired, a chemical reaction would occur, causing a sharp change in the color of the sticker. Date-specific daily stickers would be valid only on a particular date. Drivers could return unused stickers for a refund or credit. Non-date-specific daily stickers would be valid for use on any day. Complex combinations of numbers, colors, shapes, and mutilation patterns would be required to decrease the potential for the fraudulent reuse of this type of sticker. The design complexities might be reduced by limiting the use of a given set of stickers to a particular month. Unused stickers could be credited toward the purchase of another month's allotment.

Once the type of license has been selected, distribution methods must be devised. Annual stickers, which could be handled like license plates or inspection stickers, would require the smallest distribution network; daily stickers, the broadest. Like lottery tickets, daily stickers could be sold at retail outlets. Enforcement officers could also carry daily stickers to sell to out of town motorists who may not be aware of the license requirement.

Disseminating information about the program prior to its actual implementation presents a major challenge. Advertising in the media and informational displays at service stations and parking facilities would be necessary. Visitors to the city who may not know about the license requirement should be offered a chance to purchase a daily sticker before a fine is levied.

Strict enforcement is the key to the success of a licensing scheme. In Singapore, the nature of the program and the geography of the city have made effective enforcement possible. Cars must display a license only when they enter the CBD during the morning rush hour. No licenses are required to travel in the CBD at other times or within the CBD at any time. Monitoring the traffic flow to detect violators is relatively easy because only twenty-two roads lead into the CBD. Police, stationed at entrances, do not impede traffic by stopping violators. License plate numbers are noted and violators receive a ticket ($20) in the mail. In cities with a multitude of entry points or under more complex licensing schemes (i.e., congestion fees and license requirements that vary from zone to zone and time to time), enforcement would probably be difficult if not impossible. Providing stiff penalties for violators and creating a special enforcement corps (perhaps including parking facility operators and attendants) to assist the police may remove some of the incentive to circumvent the license requirement.

Although congestion pricing programs may not be politically feasible at the present time, a carefully drafted supplemental license program should not have too much trouble surviving judicial review.[82] Municipalities possess broad authority under their police powers to regulate the use of the public streets. Many cities presently have ordinances and regulations designed to reduce traffic congestion by prescribing rules for stopping, standing, and parking during peak hours. A congestion pricing program involving special licenses is a big, but logical and consistent, step beyond the existing ordinances. The provision in the federal highway law prohibiting "tolls" on federally aided highways should not prevent the implementation of a license scheme. Courts have construed "toll" narrowly, limiting its meaning to fees collected at the time and place of passage. The congestion fees charged for area licenses are akin to user charges (e.g., gasoline taxes), which courts have held do not violate the prohibition on tolls.[83]

In conclusion, congestion pricing strategies offer a variety of benefits that more than outweigh the costs imposed, even for the recalcitrant commuter who refuses to give up the luxury of driving to work alone. Congestion pricing policies should reduce urban motor vehicle congestion, air and noise pollution, and energy consumption. If, as some fear, congestion pricing has regressive income distributional consequences,[84] the revenue from congestion fees can be used to alleviate the adverse impact of congestion pricing on poor people. Direct payments to the poor is one way to provide relief; reduction in property and sales taxes, admittedly regressive levies that often finance some highway construction and maintenance, is another. In addition to these options, the revenue produced by congestion pricing can be devoted to financing service improvements in transportation alternatives to the automobile.

Despite all of these potential benefits, the congestion pricing approaches surveyed in this section may constitute too drastic a change from the status quo to be politically feasible under existing circumstances. Americans have demonstrated a remarkable tolerance for urban motor vehicle congestion and its unpleasant, unhealthy side effects, and they are too attached to their automobiles to accept the limitations that any of these congestion pricing schemes may impose. A less drastic automobile restraint policy—parking taxes—may achieve many of the same objectives as congestion pricing while perhaps being slightly more politically feasible than an area license program.

PARKING TAXES

In the days before air pollution and energy conservation became important concerns, as the automobile ascended to its preeminent position as this country's primary means of transportation, a need arose for places to store the cars when they were not on the highways. The public and private sectors responded to this demand by providing an abundant supply of low cost or free parking spaces. For some time, however, this parking situation has, along with the underpricing of public resources, exacerbated the problems of urban motor vehicle congestion. Today, the plethora of cheap parking spaces also impedes efforts to encourage the public to reduce automobile use in order to save energy.

During the last five years, the federal government and numerous local governments have studied parking management programs as one means for combating automobile air pollution and energy consumption.[85] Ideally, a comprehensive parking management program would consist of policies affecting both parking market conditions—abundant supply and low cost—that encourage automobile use. For example, measures to restrict or reduce the supply of parking spaces include:

- Amending zoning laws that require builders to construct off-street parking facilities adjacent to residential and commercial buildings;

- Limiting off-street commuter parking;
- Banning early morning and late afternoon on-street parking in the CBD; and
- Instituting a residential parking permit program in neighborhoods near employment and shopping districts.

Strategies to affect the price of parking include:

- Revising the rate structure of off-street parking facilities to discourage long term (i.e., commuter) parking;
- Eliminating free and subsidized parking for government and private employees so that they pay fees comparable to daily rates charged by commercial facilities; and
- Levying a parking tax or surcharge on off-street parking spaces.

Although a parking management plan, to be most effective, should include several of these complementary measures, this chapter will examine only the parking tax or surcharge strategy.[86]

All existing parking taxes have been enacted primarily to generate revenue. Among the cities that have parking taxes are New York City (a 6 percent sales tax), San Francisco (a 10 percent sales tax, although from 1970–1972 the rate was 25 percent), Washington, D.C. (a 12 percent sales tax), Pittsburgh (a 20 percent sales tax), and Chicago (a 15 cent excise tax).[87] Unfortunately, these revenue measures do not produce taxes high enough to influence a person's decision of whether to drive alone or to use some alternative means, such as a carpool or public transportation. For example, the tax in Pittsburgh on a typical all day parking fee of $3 is only 60 cents, hardly a significant disincentive to the continued reliance on the low occupancy automobile.

Before a municipality can impose any tax, the state must have delegated the taxation power to the municipality. Assuming for the sake of this discussion that the delegation of authority to levy the types of taxes discussed here is not a problem, a local government has two options for using its tax power to increase the price of parking. First, a city can increase existing, or impose new, levies on the owners-operators of commercial parking facilities. For example, a city could raise property taxes on parking lots and garages or enact a gross receipts tax on parking proceeds. The effectiveness of this approach in changing the travel behavior of automobile users depends not only upon the amount of the tax but also upon the owner-operator's passing on these increased or new taxes to customers by raising parking fees. Competitive pressures may compel the operator to absorb, rather than pass through, some or all of these increased taxes. When Pittsburgh, for example, enacted a gross receipts tax on parking fees, private parking operators faced a dilemma over whether to raise prices because their main source of competition, municipal parking facilities, was not subject to the tax.[88] Even if a parking operator does pass on the higher levies, the increased taxes, spread among all customers, may not amount individually to a significant disincentive to driving.

The second option for a municipality, again assuming no problem with the delegation of authority to tax, is to impose a parking tax or surcharge directly on the parking patron. Because this approach is more likely to result in an effective disincentive to automobile use, the remainder of this analysis of parking taxes concentrates on this alternative.

In designing a parking tax for energy conservation purposes, the first consideration is the scope of the tax, in terms both of its geographic boundaries and of the trip purposes (or class of parking patrons) at which it is aimed. In approaching this question, it is important to remember the object of the tax—to encourage people to drive their cars less and to shift to public transportation and paratransit services such as carpools. In other words, the goal is to save energy without significantly restricting personal mobility. Ideally, from an energy conservation perspective, a parking tax should be imposed throughout a metropolitan area on all people who use nonresidential, off-street parking facilities. In view of the generally inadequate transit service in suburban parts of metropolitan areas and the impracticality of forming carpools for nonwork trips, an areawide tax on all parking customers could significantly reduce personal mobility. Because transit service is best for rush hour trips to and from the CBD, and carpooling and other forms of ridesharing are most practical for work trips, mobility would not be unduly impaired by a parking tax or surcharge aimed at commuters to the CBD and other employment centers. Thus, anyone traveling into the designated area and parking his or her car during designated hours (e.g., 6:00–9:30 a.m.) would pay the parking tax or surcharge. With this design, a parking tax would, in essence, be a form of congestion pricing.

A tax of this scope has both advantages and disadvantages. Unlike a tax imposed on everyone entering the CBD at any time, a tax levied only on persons who park during the morning rush hour would not apply to shoppers. Retail merchants in the CBD would probably not oppose this type of tax as vigorously as they would a tax on all parking patrons that would threaten to divert more shoppers to suburban shopping centers with acres of free parking. Imposing the tax at all employment centers outside the CBD will reduce the danger of businesses and employment opportunities leaving the CBD to avoid a tax levied only in the CBD. On the other hand, the energy savings resulting from a parking tax may be offset if people seek to avoid the tax by having nonworking members of their families drive them to work, drop them off, and pick them up at night. In addition, some of the gasoline savings of those people who shift to public transportation will be burned up by an increase in nonwork travel made possible by the cars left home.[89]

Despite these side effects and the certain vociferous opposition by commuters and parking facility operators to a parking tax, the tax's potential for conserving gasoline, reducing urban motor vehicle congestion (particularly in combination with other strategies), and raising revenue justifies its serious consideration by municipal officials. The following paragraphs address some of the key issues that will arise in designing a parking tax.

Amount of Tax

All presently existing parking taxes, none of which are specifically designed to curb automobile use, have had relatively little impact on the number of cars parked or on the total level of traffic in the cities where they are imposed. Not even the 25 percent tax levied in San Francisco from 1970–1972 had much impact on these overall indicators of effectiveness, but commuters were much more sensitive to the tax than shoppers.[90] The failure of existing taxes results largely from their low level. Few people are likely to change their driving habits to save 50 to 75 cents a day.

Although stiff parking taxes have never actually been imposed, several studies of the potential impact of high taxes suggest promising results. A 1975 study of the Washington, D.C., area, prepared by the Metropolitan Washington (D.C.) Council of Governments, found that imposing either a $1 or $2 surcharge on all long term off-street parking spaces could produce significant changes in travel patterns.[91] The study, which assumed that all free and subsidized commuter spaces would be required to charge the prevailing daily commercial parking fee in addition to the surcharge, found that a $1 surcharge would reduce automobile VMT by 8 percent and increase home-to-work transit trips by 8 percent, while the $2 tax would produce a 15 percent VMT reduction and a 14 percent increase in transit trips by commuters.[92] Another analysis of the impact of parking surcharges in the Washington, D.C., area, conducted by Cambridge Systematics, Inc., estimated similar increases in transit use, but much smaller reductions in automobile VMT.[93] A 1978 study by the Urban Institute estimated that the imposition of a $2 parking surcharge during the morning rush hour at off-street parking facilities in the core area of Madison, Wisconsin, would cause nearly one-third of the parkers who drive to work alone to shift to transit or shared ride modes.[94] These reports suggest that stiff surcharges will significantly affect the travel behavior of automobile commuters.

Another factor that will be important to the effectiveness of a parking surcharge is the frequency with which it is paid. Studies of travel behavior theorize that a person's perception of the travel time, travel convenience, and travel cost characteristics of alternative modes of transportation will determine which mode (e.g., automobile or transit) he or she will choose.[95] In calculating the cost of a particular trip, the driver, according to theory, often excludes marginal operating costs (i.e., gasoline) from the computation and includes only direct out of pocket costs (i.e., tolls and parking fees) because the latter are paid more often and the driver is more aware of them. Thus, a parking tax or charge that the traveler pays every day has a greater potential for discouraging auto use than one that is paid monthly, quarterly, or annually when the parking bill, if any, is paid. Requiring daily collection of the parking surcharge may, however, impose an unreasonable burden on parking operators who do not otherwise collect daily fees.

Another factor affecting the effectiveness of stiff surcharges is whether the tax is imposed in its full amount at one time or whether it is phased in gradually. Although parking taxes might be increased in steps to coincide with transit and paratransit improvements, gradual imposition of the taxes could reduce their full effectiveness. Auto travelers might react to a gradually increasing tax as consumers react to creeping price inflation—with complaints, but without any significant or permanent changes in their buying habits. If adequate transit service exists, delaying the effective imposition of the tax for a month or two after its enactment should give automobile commuters who wish to save the money plenty of time to arrange other means of travel.

Type of Tax

A municipality can choose to impose a parking tax either as a sales tax (i.e., a percentage of the parking fee) or as a surcharge (i.e., a flat rate for the use and privilege of parking in the parking space). For the tax proposed in this section—a levy on commuters to the CBD and other employment centers—a flat rate surcharge is preferable to the percentage tax.

First, in order for a percentage tax to provide the level of tax effective in changing travel behavior, $1 or $2, a municipality would probably need to enact a percentage tax rate of as much as 50 or 100 percent. Although having the same effect as a 100 percent tax, a $2 surcharge may be perceived as more reasonable.

Second, a more substantive shortcoming of the percentage tax approach is that the disincentive effect it creates would vary according to the parking fee. Thus, a commuter to the CBD may pay twice the tax that a commuter to another employment center pays. Since the objective of the parking tax proposed in this section is to impose an equal disincentive on all commuter trips, a uniform surcharge is preferable to a percentage tax.

Finally, a percentage tax, as a sales tax based upon a transaction price, would affect only those parking transactions for which the parking patron pays a consideration or fee. Employees who park free of charge in spaces provided by their employers will escape the tax entirely. In addition, employers who charge employees a nominal parking fee may drop the fee in order to placate their employees and avoid the tax collection and bookkeeping burdens that the tax would impose. Thus, the percentage parking tax provides only partial coverage and could possibly encourage the conversion of low cost spaces to no cost spaces to escape the tax burdens.

The surcharge, on the other hand, could be designed so that it is triggered by the *activity* of occupying a parking space rather than by the *rental* of a space for a fee. Unlike a percentage tax, therefore, a flat rate surcharge could theoretically be imposed on *all* parking patrons, even those who enjoy free parking privileges. Considering whether or not to impose the tax on all parking spaces occupied in designated areas during the designated hours raises the issue of what tax exemptions, if any, to create.

Exemptions

At a minimum, the parking tax ordinance should exempt handicapped persons from the tax since they may have no choice of using another mode of transportation if their conditions require them to use specially equipped vehicles. Exempting carpools from the parking tax (i.e., cars with three or more persons would not have to pay the tax) would undoubtedly be popular, but administering this exemption could present substantial problems. On the one hand, a bona fide carpool may drop off passengers before arriving at the parking facility and thus not qualify for the exemption. On the other hand, a noncarpooler may pick up people before entering the garage and thus fraudulently obtain an exemption. Registering carpools and providing them with a tax exemption card would perhaps eliminate these problems.[96]

In addition to considering whether to exempt certain classes of commuters, the municipality should also consider whether to exempt certain types of parking facilities from taxation. Several of the existing parking taxes specifically exempt on-street metered parking and off-street parking at service stations and repair garages, bus and train depots, hotels, hospitals, and residential buildings or complexes. These exemptions seem reasonable, but are subject to abuse if the operators of these garages and lots allow commuters to park there.

Advertising Ban

Because the effectiveness of a parking tax or charge depends on the auto traveler's perception of it, a municipality should consider forbidding parking operators from advertising that they will absorb the tax or refund the tax to the patron.[97] In addition to this ban, the parking tax ordinance should require parking facility operators to erect prominent signs displaying the amount of the parking fee (if one is charged), the amount of the tax, and the total parking charge for persons subject to the tax.

Disposition of Revenue

A parking tax or surcharge, even as auto commuters switch to other modes, should generate substantial revenue. Earmarking these tax dollars for transit and paratransit service improvements may temper the political opposition to a parking surcharge.

The administrative and enforcement burdens and costs arising from a parking surcharge will vary according to the scope of the surcharge. Imposing the tax at parking facilities that collect daily parking fees and already collect and remit other taxes, should not impose any substantial additional burdens. Extending the tax to persons who pay for their parking on a monthly, quarterly, or annual basis or who park free of charge and requiring daily collection of the tax will probably create significant costs and burdens for the operators of these types of facilities and for the local government agency administering the tax. These burdens certainly will not enhance the tax's popularity.

A substantial parking surcharge, like any other form of congestion pricing, will encounter intense political opposition and possibly legal challenges from groups that believe the tax will adversely affect their interests. Commuters and parking facility owners-operators will be the primary opponents.

Commuters will be most vociferously opposed to a significant parking surcharge. A 1973 public opinion survey found that imposing taxes on all day downtown parking was the least popular energy conservation strategy considered; it was substantially less popular than higher gasoline taxes and gasoline rationing.[98] The EPA proposals for parking surcharges as part of the Clean Air Act Transportation Control Plans resulted in so much opposition that Congress prohibited their imposition.[99] In view of these attitudes and experiences, it is doubtful that parking surcharges will meet an enthusiastic reception.

If a parking surcharge on commuters to the CBD and other employment centers with adequate alternative means of transportation is enacted despite public opposition, it is certain to face court challenge. To enhance its chances for surviving judicial scrutiny, the surcharge ordinance should be drafted as a revenue-raising measure (e.g., to finance transit and paratransit service) with its regulatory objectives (i.e., changing travel behavior to reduce air pollution, congestion, and gasoline consumption) as secondary or incidental considerations. Next, the municipality should be prepared to meet an equal protection challenge alleging that no rational basis exists for taxing commuters and not other parking patrons. In upholding the Pittsburgh gross receipts parking tax, the U.S. Supreme Court stated that "[t]he city was constitutionally entitled to put the automobile parker to the choice of using other transportation or paying the increased tax."[100] The question for the parking surcharge proposed in this chapter is whether a city may constitutionally force the choice upon automobile commuters but not other parking patrons. The distinction is probably constitutionally permissible because it is rationally related to legitimate governmental concerns—reducing air pollution, urban motor vehicle congestion, and energy consumption.[101] As long as reasonably adequate alternative means of transportation exist so that the tax does not unduly impair the fundamental right to travel of those subject to the tax, the distinction between commuters and other parking patrons will probably not be held to violate equal protection principles.

Parking facility owners-operators will probably complain that a parking surcharge will adversely affect their businesses. A study of the impact of the 25 percent ad valorem tax in San Francisco, imposed from 1970–1972, found that the tax did substantially reduce the revenue, and hence the profits, of San Francisco parking operators.[102] As the Supreme Court held in *City of Pittsburgh v. Alco Parking Corp.*, however, a lawfully enacted tax that impairs the profitability of a business (or even destroys the business) does not, on that ground alone, unconstitutionally deny the businessman's due process rights.[103]

In conclusion, local government officials who are attempting to design an effective, politically feasible parking surcharge must keep two considerations foremost in their minds. First, it is essential that a municipality not impose a substantial parking surcharge unless reasonable, adequate transportation alternatives exist. The purpose of the tax is to force people to make a choice between paying society for polluting the air, consuming nonrenewable energy, and creating traffic congestion on the one hand and, on the other, using their cars less and switching to alternative transportation modes such as transit and paratransit. The surcharge cannot force this choice if no reasonable alternative exists to which people can turn to avoid paying the tax. The existence of adequate transit and/or paratransit service to and within the area in which the surcharge is applied is, therefore, a vitally important prerequisite to the imposition of a surcharge.

Second, municipalities that decide to attack the energy consumption of automobiles by implementing a parking management program cannot expect to achieve dramatic results with a program consisting of a single strategy—a parking tax. For example, without a complementary strategy to limit commuter parking in residential neighborhoods adjacent to employment centers, many commuters will attempt to escape the tax at commercial facilities by parking for free in residential areas near their offices. A parking management plan that integrates several of the strategies designed to deal with the price and supply aspects of the various types of parking spaces will be far more effective than a parking tax alone.

CONCLUSION

The choices Americans make when buying automobiles and the habits they follow in using them have a significant effect on the success of national energy conservation efforts. This chapter has examined five economic disincentives that states and localities can enact to encourage people to make choices and develop habits that save energy.

Two of the disincentives will reinforce federal efforts to improve fuel economy. If not preempted by the federal gas guzzler tax, a state efficiency tax-rebate program for new cars will establish new, stronger incentives for people to purchase efficient cars. Revising automobile registration fees so that the fees vary according to vehicle weight will provide an annual reminder that it pays to own efficient cars. Although neither strategy will substantially cut gasoline consumption in the short run, both, especially in combination, should be effective in the long run. Moreover, because neither would restrict individual mobility, their chances of surviving the political gauntlet are greater than those of the other three disincentives discussed in this chapter.

The other economic disincentives are designed to reduce automobile VMT by encouraging people to switch from their cars to more energy-efficient modes of transportation, such as transit and paratransit, when these alternatives are available. Sharply higher gasoline taxes (e.g., 50 cents per gallon), which studies suggest would result in large gasoline savings only after several years, would impair mobility by indiscriminantly discouraging all trips, including those for which no alternative means of transportation exist. Modest gasoline tax increases, however, would not have as devastating an effect on mobility and would raise substantial revenue. With the modification of constitutional and statutory prohibitions on the use of gasoline tax revenues for nonhighway purposes, the money could be used to finance energy-efficient transportation alternatives to the automobile.

Unlike gasoline taxes, congestion pricing programs and parking taxes can be imposed in a way that discourages those trips for which reasonable and adequate transportation alternatives exist. If good transit or paratransit services are available, these strategies should convince many people to leave their cars at home.

These five economic disincentives are not, by any means, the only effective strategies for dealing with the contribution of the automobile to the nation's energy problem. Another book in this series, *Saving Energy in Urban Transportation*, by Durwood J. Zaelke, Jr., explores an entire range of policies designed to reduce the nation's reliance on the automobile.

These policies and the economic disincentives discussed here, however, face an enormous obstacle—Americans' still strong love affairs with their cars. The depth of these feelings are suggested by the vituperative language of a letter to the editor of the *Washington Post*, in the Fall of 1978.[104] The letter writer, an automobile commuter, characterized efforts to convince people to use public transportation instead of their cars as an "insane cause" and described those officials and citizens who support these efforts as "a small minority, but including the best thinkers of the 18th century."[105] As long as attitudes of this kind prevail, and until the love affair wanes, the odds in favor of the enactment and implementation of the economic disincentives analyzed in this chapter are, unfortunately, not very encouraging.

NOTES TO CHAPTER 4

1. P. L. 94–163, codified in part at 42 U.S.C. §§ 6201 et seq. and 15 U.S.C. §§ 2001–2012.

2. 42 U.S.C. § 6261(b)(3)(A) and proposed energy conservation contingency plans at 41 Fed. Reg. 21908 (May 28, 1976).

3. 42 U.S.C. § 6322(d)(3).

4. 15 U.S.C. §§ 2001–2012.

5. 49 C.F.R. § 531.5. According to the Department of Transportation, these standards are "intended to result in the consumption of approximately 41 billion fewer gallons of gasoline . . . over the life of the vehicles manufactured in 1981–84 than would be the case if the average fuel economy of new passenger automobiles remained at the level of the 1980 fuel economy standard, 20.0 mpg." 42 Fed. Reg. 33534 (June 30, 1977). For supplementary information on these standards, see 42 Fed. Reg. 33534–552 (June 30, 1977).

6. C. Difiglio, "A Comparison of Mandatory Fuel Economy Standards and Automobile Excise Taxes" (preliminary draft; presented at the Joint Annual Meeting of the Operations Research Society of America and the Institute of Management Sciences, March 31, 1976), Table 2, p. 41.

7. "EPA Reports Gas Mileage Exceeds '78 U.S. Standards," *Washington Post*, September 23, 1976, p. A–1.

8. 15 U.S.C. § 2008.

9. National Energy Conservation Policy Act of 1978, P. L. 95–619, § 402, codified at 15 U.S.C. § 2008(d).

10. Many people began to ignore EPA's ratings after they discovered that their actual mileage was much lower than EPA estimates. *See* Patrick O'Donnell, "Auto Buyers Disgruntled by EPA's Gas-Mileage Estimates," *Washington Post*, September 3, 1977, p. A–3. As a result of the public's complaints about the mileage figures, EPA adopted a new ratings system for the 1979 model year. *See infra* note 22.

11. 15 U.S.C. § 2006.

12. *See, e.g.*, "Back to 'More Car per Car'," *Time*, June 14, 1976, p. 53; and Owen Ullmann, "Big-Car Demand Pushes Sales to 12-Year Record," *Washington Post*, February 16, 1977, p. D–11.

13. Executive Office of the President, Energy Policy and Planning, *The National Energy Plan* (Washington, D.C.: U.S. Government Printing Office, 1977), xv; H.R. 6831 (95th Congress), §§ 1201–1204, 1221–1223.

14. Energy Tax Act of 1978, P. L. 95–618, § 201, Internal Revenue Code (I.R.C.) § 4064.

15. 42 U.S.C. § 6326(2).

16. 42 U.S.C. § 6322(a)(1).

17. Paratransit modes include hire and drive services (e.g., daily or short term rental cars), hail or phone services (e.g., taxicabs, dial a ride buses, and jitneys), and prearranged ridesharing services (e.g., carpools, vanpools, and subscription buses). In other words, paratransit includes practically everything except the conventional use of the private automobile and conventional transit. *See generally* R.F. Kirby et al., *Para-Transit—Neglected Options for Urban Mobility*, (Washington, D.C.: Urban Institute, 1974); and D.J. Zaelke, Jr., *Saving Energy in Urban Transportation*, (Cambridge, Massachusetts: Ballinger Publishing Co., 1979).

18. Federal proposals in the 94th Congress (1975–1977) included the Fisher amendment to H.R. 6860 121 Cong. Rec. 15153–54 (1975) and S. 973, and proposals in the 95th Congress (1977–1979) included H.R. 820. *See also* T.J. Gallagher, Jr., "An Automobile Fuel Consumption Efficiency Tax: A Sumptuary Response to the 'Efficiency Crisis'," 7 *Rut.-Cam. L. J.* 1 (1975).

State proposals include Arizona—H.B. 2148 (1976); Connecticut—Proposed Bill 6000 (1975); Connecticut—Committee Bill 5452 (1976); Colorado—H.B. 1244 (1976); Maryland—H.B. 702 (1975); Maryland—H.B. 82 (1976); Massachusetts—S.B. 1111 (1976); Minnesota—H.F. 1718 (1975); and Minnesota— H.F. 1171 (1977).

19. Energy Tax Act of 1978, P.L. 95–618, § 201, I.R.C. § 4064.

20. R. Landis, *The Effect of Automotive Fuel Conservation Measures on Air Pollution*, EPA–600/5–76–006, prepared for the U.S. Environmental Protection Agency by Charles River Associates, Inc. (Washington, D.C.: Environmental Protection Agency, September 1976), p. 10.

21. Colorado—H.B. 1244 (1976).

22. Beginning with the 1979 model year, EPA changed its fuel-economy-rating system. *See supra* note 10. Instead of publishing separate figures for city and highway fuel economy and combining these figures to obtain an overall fuel economy rating, EPA began to publish a single fuel economy figure that is roughly equivalent to the city figure of past years.

For purposes of determining compliance with the EPCA fuel economy standards and applicability of the gas guzzler tax, however, fuel economy is required to be calculated according to the old method that derived a single fuel economy rating by combining the figure for city driving (weighted 55 percent) and the figure for highway driving (weighted 45 percent). 15 U.S.C. § 2003(d)(1) and I.R.C. § 4064(c)(1). Because of the preemption questions that conflicts between the federal fuel economy programs and a state tax-rebate program might raise, states should probably define fuel economy in terms that are consistent with the provisions of EPCA and the gas guzzler tax.

23. Arizona—H.B. 2148 (1976).

24. Maryland—H.B. 82 (1976).

25. The ratings for 1979 models reflect the fuel economy achieved by the cars under city driving test conditions. *See supra* note 22.

26. Maryland—H.B. 702 (1975).

27. This proposal was developed by Christopher Wasintynski, chairman, National Transportation Committee of the Sierra Club. (Letter to Michael McCloskey and Ellen Winchester, November 13, 1976).

28. Arizona—H.B. 2148 (1976); and Connecticut—Committee Bill 5452 (1976).

29. H.R. 6831 (95th Congress), §§ 1201–1204; and the Energy Tax Act of 1978, P. L. 95–618, § 201, I.R.C. § 4064(a).

In December 1978, Ford Motor Company raised the prices of its big cars to discourage sales of these models. Ford took this action because big car sales were jeopardizing its chances of meeting the EPCA standard for the 1979 model year. *Washington Post*, December 13, 1978, p. E–7.

30. Colorado—H.R. 1244 (1976); Connecticut—Proposed Bill 6000 (1975); and H.R. 6831 (95th Congress), §§ 1201–1204.

31. *See, e.g.,* "Cadillac Mystique: 'Looking Good'," *Washington Post*, February 23, 1977, p. A–1.

32. Maryland—H.B. 702 (1975).

33. The relevant provision in EPCA states that whenever the federal labeling standards apply to an automobile, "no State or political subdivision of a State shall have authority to adopt or enforce any law or regulation with respect to *the disclosure of fuel economy of such automobile . . . if such law or regulation is not identical with such [labeling] requirement*" (emphasis added). 15 U.S.C. § 2009(b).

34. Maryland—H.B. 82 (1976).

35. Several analyses, however, were conducted to determine the energy impact of President Carter's tax-rebate proposal. *See* Congressional Budget Office, *President Carter's Energy Proposals: A Perspective*, Staff Working Paper, 2nd ed. (Washington, D.C.: U.S. Government Printing Office, 1977), pp. 53–79; and "Energy Tax Proposals Relating to Transportation," prepared by the Joint Committee on Taxation for the Committee on Ways and Means, U.S. House of Representatives (Washington, D.C.: U.S. Government Printing Office, 1977), pp. 15–29. *See also* Landis, *supra* note 20, Table S–3 at 18. (This study estimates the effect on gasoline consumption—virtually no savings in the short run and modest savings in the long run—of federal excise taxes on new cars of either $50, $100, or $200 per mile per gallon for each mile per gallon less than twenty.)

36. Federal Highway Administration, *Highway Statistics (1974)* (Washington, D.C.: U.S. Government Printing Office, 1975), Table MV–103, a summary of state motor vehicle registration fee schedules in effect as of January 1, 1976.

37. D.C. Code § 40–103(b)(1) (as amended June 15, 1976). In 1977, however, the registration fees were reduced. The fee schedule applicable after September 30, 1977, is $35 for Class I vehicles, $42 for Class II, $68 for Class III and $76 for Class IV. D.C. Code § 40–103(b)(1) (as amended April 19, 1977).

38. Enacting registration fees under the tax power allows states to impose fees that are graduated and that generate "excess" revenue. *See generally* 7 Am. Jur. 2d *Automobiles and Highway Traffic* § 63 (1963). *See also* the discussion of the differences between fees enacted under the police power and fees enacted under the tax power, *supra* pp. 34–35.

39. *See* Federal Highway Administration, *Highway Statistics (1975)* (Washington, D.C.: U.S. Government Printing Office, 1976), Table MV–106, a summary of provisions (effective as of January 1, 1977) governing the disposition of state motor vehicle and motor carrier receipts, including registration fees.

40. *Highway Statistics (1974)*, *supra* note 36, Table MF—21.

41. *Highway Statistics (1975)*, *supra* note 39, Table MF—21.

42. Federal Highway Administration, *Highway Statistics (1976)* (Washington, D.C.: U.S. Government Printing Office, 1977), Table MF—21. The 1977 and 1978 growth rate figures were supplied by the Office of Highway Statistics, Federal Highway Administration.

43. Landis, *supra* note 20 at 18.

44. In EPCA, Congress provided for the preparation of a rationing contingency plan to be used in the event of "a severe energy supply interruption." 42 U.S.C. § 6261(b)(3)(A).

45. *Highway Statistics (1976)*, *supra* note 42 at Table MF—1 (state gasoline taxes as of December 31, 1976).

46. Organisation for Economic Cooperation and Development, *Energy Conservation in the International Energy Agency—1976 Review* (Paris, 1976).

47. Harris Survey, "Americans Concerned About Energy" (February 14, 1977).

48. Some of the studies of the elasticity of demand for gasoline are summarized and discussed in S. Wildhorn et al., *How to Save Gasoline: Public Policy Alternatives For the Automobile*, R—1560—NSF, prepared for the National Science Foundation (Santa Monica, California: Rand Corporation, 1974), Table 6—3, p. 64; R.G. McGillivray, "Gasoline Use by Automobiles," Working Paper 1216—2 (Washington, D.C.: Urban Institute, December 1974), pp. 5—8; and R.G. McGillivray, "Automobile Gasoline Conservation," Paper 708—01 (Washington, D.C.: Urban Institute, April 1976), pp. 15—21.

49. *See* Data Resources, Inc., *A Study of Quarterly Demand for Gasoline and Impacts of Alternative Gasoline Taxes*, prepared for the Environmental Protection Agency and the Council on Environmental Quality (Lexington, Massachusetts: Data Resources, Inc., December 1973), estimating a short run elasticity of −0.075; Wildhorn, *supra* note 48, estimating a short run elasticity of −0.10 to −0.18 using one model and −0.83 using another; McGillivray, "Automobile Gasoline Conservation," *supra* note 48, estimating a short run elasticity of −0.10 to −0.30; J. Ramsey, A. Rasche, and B. Allen, "An Analysis of the Private and Commercial Demand for Gasoline" (unpublished paper, February 18, 1974), referred to in Wildhorn, *supra* note 48 at 64, estimating a short run elasticity of −0.77; C. Chamberlain, "Models of Gasoline Demand" (unpublished paper, Fall 1973), referred to in Wildhorn, *supra* note 48 at 64, estimating a short run elasticity of −0.06; Landis, *supra* note 20, estimating a short run elasticity of −0.16.

50. *See* Data Resources, Inc., *supra* note 49, estimating a long run elasticity of −0.24; C. Chamberlain, "Models of Gasoline Demand," *supra* note 49, estimating a long run elasticity of −0.07; Wildhorn, *supra* note 48, estimating a long run elasticity of −0.92; Landis, *supra* note 20, estimating a long run elasticity of −0.59 to −0.65.

51. J. Suhrbier et al., *Carpool Incentives: Analysis of Transportation and Energy Impacts*, FEA/D—76/391, prepared for the Federal Energy Administration by Cambridge Systematics, Inc. (Washington, D.C.: U.S. Government Printing Office, 1976), Table 16, p. 58.

52. Wildhorn, *supra* note 48.

53. *Id.*, Table S—1 at ix.

54. *Id.*, Table 6—4 at 66.

55. Landis, *supra* note 20, Table S—3 at 18.

56. R.L. Schaeffer et al., "The Immediate Impact of Gasoline Shortages on Urban Travel Behavior," prepared for Federal Highway Administration, January 1975, reported in *Carpool Incentives: An Evaluation of Operational Experience*, Federal Energy Administration, Energy Conservation Paper No. 44 (March 1976), pp. 139—41.

57. *See* Landis, *supra* note 20 at 170—72; and Data Resources, Inc., *supra* note 49 at IV.34—37. *See also* the studies listed in J.P. Stucker, "The Distributional Implications of a Gasoline Tax," prepared for 1975 meeting of the Western Economic Association, San Diego, California, June 27, 1975, a paper that disputes the conclusion that gasoline tax increases are not regressive.

58. *See Highway Statistics (1975), supra* note 39 at Table MF—106, a summary of provisions governing the disposition of state motor fuel tax receipts as of January 1, 1977.

59. Mass. Const. amendment article 104 (1974).

60. In 1976 the Michigan legislature in Public Act 297 established the General Transportation Fund to finance public transportation and appropriated 0.05 cent of the gasoline tax to the fund. In Advisory Opinion Re 1976 PA 295 and 1976 PA 297, 401 Mich. 686, 259 N.W.2d 129 (1977), the Michigan Supreme Court upheld the constitutionality of this action. According to the court, the provision in Michigan's constitution dedicating gasoline taxes "exclusively for highway purposes as defined by law" was not violated when the legislature in PA 297 defined highway purposes to include public transportation services.

61. Chapter 770, S.L. 1975. The law required unanimous favorable action by all the affected localities, and because one of them balked, the tax was never imposed.

62. In August 1978, Missouri voters overwhelmingly rejected a proposed 3 cents per gallon increase in gasoline taxes that proponents argued was needed for highway repair. "Primary Day Kind to Incumbents," *Washington Post*, August 10, 1978, p. A—4.

63. W. Elliott, "Hidden Costs, Hidden Subsidies: The Case for Road Use Charges in Los Angeles," Working Paper, February 14, 1975 (Institute of State and Local Government, Claremont Men's College, Claremont, California). *See also* W. Elliott, "VMT Disincentive Choices for Los Angeles Basin," undated paper; and W. Elliott, "Giving the Plan a Bottom Line: Suggestions for adding cost-benefit comparisons to the California Transportation Plan," revised version, May 12, 1976.

64. W. Elliott, "Hidden Costs, Hidden Subsidies," *supra* note 63 at 11.

65. *See, e.g.*, K. Bhatt, "What Can We Do About Urban Traffic Congestion? A Pricing Approach," Working Paper 5032—03—1 (Washington, D.C.: Urban Institute, February 1975); K. Bhatt, "Road Pricing Technologies: A Survey," Paper 1212—11 (Washington, D.C.: Urban Institute, August 1974); D. Kulash, "Congestion Pricing: A Research Summary," Paper 1212—99 (Washington, D.C.: Urban Institute, July 1974); R.G. McGillivray, "On Road Congestion Theory," Paper 1212—8—1 (Washington, D.C.: Urban Institute, March 1974);

A.A. Walters, "The Theory and Measurement of Private and Social Cost of Highway Congestion," 29 *Econometrica* 676 (1961).

66. *See generally Urban Transportation Pricing Alternatives*, a collection of papers presented at a conference on May 14–17, 1976, at Easton, Md., review draft (Washington, D.C.: Transportation Research Board, 1976); and the series of Urban Institute papers on congestion pricing resulting from research supported by the National Science Foundation and the Urban Mass Transportation Administration of the U.S. Department of Transportation.

Evidence of government interest in congestion pricing is provided by U.S. Department of Transportation regulations, 23 C.F.R. §§ 450.100–.122, and Appendix A, 40 Fed. Reg. 42976 (September 17, 1975), suggesting congestion pricing as a category of action that state transportation planners should consider including in their Transportation Systems Management plan, and by the Urban Mass Transportation Administration's intention to conduct demonstrations of congestion pricing under its Service and Methods Demonstration Program.

67. Berkeley, California, and Madison, Wisconsin, however, expressed interest in participating in the Urban Mass Transportation Administration's Service and Methods Demonstration Program for congestion pricing, and the Urban Institute conducted studies of how congestion pricing policies would work in these cities and what impacts they would have. *See* M.D. Cheslow, "A Road Pricing and Transit Improvement Program in Berkeley, California: A Preliminary Analysis," Paper 5050–3–6 (Washington, D.C.: Urban Institute, September 1978); F. Spielberg, "Transportation Improvements in Madison, Wisconsin: Preliminary Analysis of Pricing Programs for Roads and Parking in Conjunction with Transit Changes," Paper 5050–3–7 (Washington, D.C.: Urban Institute, November 1978).

68. *See* A.D. May, "The London Supplementary Licensing Study," in *Urban Transportation Pricing Alternatives, supra* note 66.

69. *See* P.L. Watson and E.P. Holland, "Congestion Pricing—The Example in Singapore," in *Urban Transportation Pricing Alternatives, supra* note 66; "Singapore Is Waging a War Against Cars—And It's Winning," *Wall Street Journal*, November 5, 1976, p. 1; "Pushing an Auto-Ban in Downtown Singapore," *Washington Post*, January 5, 1977, p. D–6; and "An Anti-Car Campaign," *Washington Post*, September 12, 1976, p. L–22.

70. The information on toll booth collection systems was found in Bhatt, "What Can We Do About Urban Traffic Congestion?" *supra* note 65 at 22–24; Bhatt, "Road Pricing Technologies," *supra* note 65 at 5–8; and Kulash, *supra* note 65 at 6–7.

71. *See* Bhatt, "Road Pricing Technologies," *supra* note 65 at 6.

72. 23 U.S.C. § 301.

73. In 1968 a survey of the twenty largest U.S. cities found that the percentage of city road miles built with federal aid ranged from 5.3 percent in Los Angeles to 23.2 percent in Cleveland. R.J. Coit, "Legal Issues Surrounding Roadway Pricing on City Streets and Bridges," Paper 1212–6 (Washington, D.C.: Urban Institute, July 1974), pp. 9–10.

74. *See* Coit, *supra* note 73, for an explanation of the tangled web.

75. The information on AVI systems was found in Bhatt, "What Can We Do About Urban Traffic Congestion?" *supra* note 65 at 24–25; Bhatt, "Road Pricing Technologies," *supra* note 65 at 13–17; and R.S. Foote, "Collection Problems and the Promise of Automatic Vehicle Identification," in *Problems in Implementing Roadway Pricing*, Transportation Research Record No. 494 (Washington, D.C.: Transportation Research Board, 1974).

76. Kulash, *supra* note 65 at 9.

77. Bhatt, "Road Pricing Technologies," *supra* note 65 at 15–16.

78. The information on automobile-mounted meters was found in Bhatt, "What Can We Do About Urban Traffic Congestion?" *supra* note 65 at 17–20.

79. The information on supplemental and special area licenses was found in Bhatt, "What Can We Do About Urban Traffic Congestion?" *supra* note 65 at 23–24; Bhatt, "Road Pricing Technologies," *supra* note 65 at 8–12; K. Bhatt and M. Beesky, "Planning and Implementing a Congestion Pricing and Transportation Improvement Package," Working Paper 5050–3–1(3) (Washington, D.C.: Urban Institute, February 1976); and K. Bhatt, J. Eigen, and T. Higgins, "Implementation Procedures for Pricing Congested Roads," Working Paper 5032–3–3(2) (Washington, D.C.: Urban Institute, February 1976).

80. *See, e.g.*, Bhatt, "What Can We Do About Urban Traffic Congestion?" *supra* note 65 at 13–15; and R.G. McGillivray, "Estimates of Optimal Congestion Tolls," Paper 1212–8–2 (Washington, D.C.: Urban Institute, July 1974). *See also* Cheslow, *supra* note 67, and Spielberg, *supra* note 67, which project the effects of charging $2 for daily area licenses in Berkeley, California, and Madison, Wisconsin.

81. *See* Bhatt, Eigen, and Higgins, *supra* note 79 at 18–26.

82. *See generally* Coit, *supra* note 73, and J.J. Bosley and M.B. Schaller, "Legal Considerations in Urban Transportation Pricing," in *Urban Transportation Pricing Alternatives, supra* note 66.

83. *See* Coit, *supra* note 73 at 5–10. Coit, however, points out a distinction between gasoline taxes and congestion fees. Gasoline taxes have traditionally been revenue measures used by the states to finance their road maintenance obligations. Congestion fees, however, have an entirely different purpose—to regulate road use. In view of the narrow judicial interpretation given to "tolls," it is unlikely that courts will seize upon this distinction in the purposes of the two types of user fees to invalidate a licensing scheme on the ground that it levies a prohibited toll on federally aided highways.

84. *See, e.g.*, A. Walters, "Distributional Effects of a Congestion Tax," Working Paper 1212–7 (Washington, D.C.: Urban Institute, September 1973); D.J. Kulash, "Income-Distributional Consequences of Roadway Pricing," Paper 1212–12 (Washington, D.C.: Urban Institute, July 1974).

85. *See, e.g.*, Clean Air Act Transportation Control Plans promulgated by the U.S. Environmental Protection Agency for the District of Columbia, Los Angeles, San Francisco, San Diego, and Boston, which included commuter parking surcharges; the Transportation Systems Management regulations of the U.S. Department of Transportation, 23 C.F.R. §§ 450.100–.122 Appendix A; Metropolitan Washington (D.C.) Council of Governments, National Capital Re-

gion Transportation Planning Board, *Parking Management Policies and Auto Control Zones*, Report No. DOT–OS–400045–1, Final Report, prepared for U.S. Department of Transportation (Washington, D.C.: Metropolitan Washington (D.C.) Council of Governments, June 1976); Southern California Association of Governments, *Draft Final Report of the Parking Management Planning Study* (Los Angeles: Southern California Association of Governments, January 1976); and "A Proposal for Development of a Parking Management Plan for the City of Los Angeles" (Los Angeles: Office of the Mayor, June 4, 1975).

86. *See* D.J. Zaelke, Jr., and J.W. Russell, Jr., "Energy Conservation Through Automobile Parking Management," *ECP Report* No. 6 (May 1976; reprinted in Zaelke, *supra* note 17) for an analysis of all these parking management measures.

87. N.Y. Tax Law § 1212–A (McKinney 1975) (enabling legislation); San Francisco Municipal Code, Part III, Article 9, §§ 601 et seq. (1972); D.C. Code § 47–2601.14 (a)(12)(1975); City of Pittsburgh, Ordinance No. 30 of 1973 (reprinted in Appendix 4–3); Municipal Code of Chicago, Chapter 156.1 (1971).

88. *See* City of Pittsburgh v. Alco Parking Corp., 417 U.S. 369, 372, n. 3 (1974).

89. *See* Suhrbier, *supra* note 51, Table 11 at 43.

90. D. Kulash, "Parking Taxes as Roadway Prices: A Case Study of the San Francisco Experience," Paper 1212–9 (Washington, D.C.: Urban Institute, March 1974), p. 17.

91. Metropolitan Washington (D.C.) Council of Governments, *supra* note 85.

92. *Id.*, Table 9–3 at 181 and Table 9–7 at 186. The percentage figures were derived from data in these tables.

93. Suhrbier, *supra* note 51, Table 11 at 43.

94. Spielberg, *supra* note 67 at 25.

95. *See, e.g.*, Suhrbier, *supra* note 51 at 9–20 and Appendix B; and D. Hiatt, "The Potential for Modal Shift: Urban Auto Travel to Public Transit," Working Paper No. RP–SP–20 (Cambridge, Massachusetts: Transportation Systems Center, 1973).

96. Aside from the enforcement difficulties of a carpool exemption, it can be argued that the exemption is not needed as a carpool incentive. The existence of the tax, the impact of which people can reduce by joining a carpool, is already a sufficient incentive. The impact of a $2 tax for an individual who formerly drove to work alone can be reduced to 50 cents if he rides to work with three other people. The extra incentive provided by reducing the impact to zero is probably not worth the difficulties created by the exemption.

97. *See, e.g.*, San Francisco Municipal Code, Part III, Article 9, § 605.

98. *See* Schaeffer et al., *supra* note 56 at 122–23.

99. *See* the Energy Supply and Environmental Coordination Act of 1974, P. L. 93–319, § 4(b)(2), 42 U.S.C. § 7410(c)(2)(B).

100. City of Pittsburgh v. Alco Parking Corp., 417 U.S. 369, 379 (1974).

101. *Cf.* Railway Express Agency v. New York, 336 U.S. 106 (1949) (upholding traffic regulation prohibiting all advertising on motor vehicles except for advertising of vehicle owner's business); and Lehnhausen v. Lake Shore Auto Parts, 410 U.S. 356 (1973) (upholding state constitutional provision exempting

all personal property owned by individuals from personal property taxes, but retaining taxes for personal property owned by corporations and other "non-individuals").

102. Kulash, *supra* note 90 at 22–23.

103. City of Pittsburgh v. Alco Parking Corp., 417 U.S. at 373–74, citing Magnano Co. v. Hamilton, 292 U.S. 40 (1934); and Alaska Fish Co. v. Smith, 255 U.S. 44 (1921).

104. Letter to the editor by William F. Fuchs, *Washington Post*, September 29, 1978, p. A–16.

105. *Id.*

Appendices to Chapter 4

Appendix 4-1

THE EFFECT OF THE FEDERAL GAS GUZZLER TAX ON PROPOSALS FOR STATE AUTOMOBILE EFFICIENCY TAX-REBATE PROGRAMS

Enactment of a federal gas guzzler tax as part of the National Energy Act[1] raises questions about the continuing viability of proposals for state automobile efficiency tax-rebate programs. This appendix examines an issue that bears on the legal viability of these proposals—whether the federal tax preempts state tax-rebate programs.

As part of its examination of federal constitutional limitations on the states' exercise of their police and taxing powers, Chapter 3 discusses the preemption doctrine. As outlined there, a court will strike down a state law on preemption grounds if the court finds that (1) Congress has expressly stated its intention to prohibit concurrent or supplementary state regulation of the subject matter involved, (2) the state law directly conflicts with federal law or policy, or (3) Congress has implicitly preempted state action in the field.

In passing the federal gas guzzler tax, Congress did not expressly preempt state tax-rebate programs similar to the one discussed in this chapter. Neither the Energy Tax Act of 1978 nor its legislative history contain any declaration that the federal tax is to be the exclusive means for discouraging consumers from buying inefficient cars. In the National Energy Conservation Policy Act of 1978, by comparison, Congress did expressly preempt state laws and regulations establishing energy efficiency standards for major home appliances.[2] In earlier energy legislation, the Energy Policy and Conservation Act of 1975 (EPCA), Congress expressly preempted state laws or regulations relating to automobile fuel economy standards.[3]

In deciding a preemption challenge to a state automobile efficiency tax-rebate program, a court would have little basis for holding that such a program is void because it, in operation or purpose, directly conflicts with or hinders federal law or policy.[4] State tax-rebate programs would not hinder the operation of the federal tax program any more than state income tax programs hinder the functioning of the federal income tax system. Furthermore, enactment of state tax-rebate programs would be entirely consistent with federal policies emphasizing energy conservation. More specifically, state tax-rebate programs represent an appropriate response to a 1975 congressional invitation to the states to participate in the energy conservation effort. As part of EPCA, Congress authorized federal financial and technical aid to assist states in preparing and implementing state energy conservation plans.[5] In addition to five mandatory measures that plans must contain to be eligible for assistance, states can include in their plans "any other appropriate method or program to conserve and improve efficiency in the use of energy."[6] State tax-rebate programs would not only be consistent with the federal policy, evidenced in this law, that states play a major role in encouraging energy conservation, but they would also effectively complement federal programs to improve the fuel economy of the nation's automobile fleet.

Even though Congress has not expressly preempted state tax-rebate programs, a court could conceivably find that Congress intended to preempt such programs when it passed the gas guzzler tax. Courts have found a congressional intent to preempt state laws where (1) the matter requires the national uniformity that exclusive federal regulation would provide, (2) the federal interest in the matter predominates over any state interest, or (3) the federal scheme is so pervasive as to permit the reasonable inference that Congress intended to prohibit concurrent or supplementary state action.[7]

Discouraging the sale of inefficient cars with high taxes and encouraging the sale of efficient cars with alluring rebates does not seem analogous to those matters, such as nuclear power regulation, that courts have found to require national uniformity. Similarly, there does not seem to be the type of dominant federal interest in this matter that has been held in other cases to preclude state laws on a particular subject. As for the pervasiveness of the federal scheme, an analysis of the gas guzzler tax reveals that the federal government has hardly entered the field of influencing consumers' automobile purchasing decisions, much less fully occupied it. The federal tax will have an extremely limited impact, affecting only the most egregious gas guzzlers.[8] The federal tax in no way discourages consumers from buying other relatively inefficient models (i.e., those whose fuel efficiency falls below the applicable federally mandated sales-weighted fleet fuel economic standard) or encourages them to buy efficient models.

Although a court could take a different view and arrive at the opposite conclusion, this brief review of the preemption doctrine suggests that state automobile efficiency tax-rebate programs are not barred by the federal gas guzzler tax. Consistent with federal energy policy, a state tax-rebate program, like the one discussed in this chapter, would effectively fill in the gaps left by the federal scheme.

NOTES TO APPENDIX 4-1

1. The Energy Tax Act of 1978, P. L. 95—618, § 201.

2. The National Energy Conservation Policy Act, P. L. 95—619, § 424, amending 42 U.S.C. § 6297(b).

3. The Energy Policy and Conservation Act, P. L. 94—163, § 509(a), 15 U.S.C. § 2009(a).

4. See the discussion in Chapter 4 suggesting that states use the federal definition of fuel economy and devise labeling requirements carefully to avoid creating conflicts between a state program and federal law.

5. *See* the Energy Policy and Conservation Act, P. L. 94—163, §§ 361—366, and the Energy Conservation and Production Act, P. L. 94—385, §§ 431—432, 42 U.S.C. §§ 6321—6327. *See also* the National Energy Conservation Policy Act of 1978, P. L. 95—619, §§ 621—622, which reaffirms the invitation by authorizing $100 million in federal assistance for fiscal year 1979.

6. 42 U.S.C. § 6322(d).

7. *See* the cases listed in Chapter 3 notes 33—36, *supra.*

8. The Conference Report (No. 95—1773) accompanying the Energy Tax Act of 1978 states that "the fuel efficiency standards below which a passenger automobile will be subject to tax will generally start from 4 to 6.5 miles per gallon (depending on the year involved) below the fleetwide average standards of EPCA" at 45.

Appendix 4–2

STATE OF COLORADO, HOUSE BILL 1244

A Bill for an Act Promoting the Purchase of Energy Efficient Motor Vehicles.

Introduced in the Second Regular Session, Fiftieth General Assembly, 1976. Died in Committee.

Be it enacted by the General Assembly of the State of Colorado:

SECTION 1. Article 3 of title 42, Colorado Revised Statutes 1973, as emended, is amended BY THE ADDITION OF A NEW SECTION to read:

42–3–134. *Ownership tax for energy inefficiency—payment for energy efficiency.* (1) The general assembly finds and declares that:

(a) A serious shortage of refined petroleum products is emerging in the United States;

(b) Automobiles are the single largest and most significant user of petroleum products, consuming over one billion gallons of gasoline per year in this state;

(c) Automobiles consume over thirty percent of this state's total energy consumption and the state's consumption of gasoline has been increasing rapidly;

(d) The efficient transport of passengers is vital to the economic well being of this state and the automobile is essential to this system; and

(e) The tax contained in this section is on the ownership of certain automobiles.

(2) The general assembly further finds and declares that the purpose of this section is to reduce the energy consumption of automobiles by:

(a) Increasing public awareness of the factors which influence the energy efficiency of automobiles; and

(b) Encouraging consumers to purchase automobiles which are more energy efficient.

(3) As used in this section, unless the context otherwise requires:

(a) "Automobile" means passenger-carrying motor vehicles having a seating capacity of ten or less persons and includes cars, station wagons, noncommercial and recreational vehicles, passenger vans, and pickup trucks of less than six thousand pounds empty weight. The term shall not include mobile machinery, farm tractors, self-propelled construction equipment, implements of husbandry, motor-driven cycles, motorscooters, motorcycles, school buses, truck tractors, taxi cabs, hearses, farm trucks, passenger buses for hire, authorized emergency vehicles, or stake, platform, or rack trucks.

(b) "Dealer" means any person licensed under the laws of this state to engage in the business of buying, selling, exchanging, or otherwise trading motor vehicles.

(c) "Energy efficiency" means the average number of miles traveled by an automobile per gallon of gasoline, or equivalent amount of other fuel, consumed as determined by the administrator of the federal environmental protection agency for combined city and highway driving.

(d) "New automobile" means any automobile title being transferred for the first time from a manufacturer or importer, or dealer or agent of a manufacturer or importer, which automobile had theretofore not been used and which automobile is commonly known as a "new automobile."

(4) (a) On and after July 1, 1976, and at the time of initial registration of a 1977 model automobile and any new automobile initially registered on or after January 1, 1977, there shall be made a determination of the energy efficiency of the automobile. Based upon such determination, the owner of the automobile shall either pay an ownership tax, in addition to any other tax or fee collected, or shall be entitled to a payment from the energy conservation reserve account created in paragraph (b) of this subsection (4). The amount of the tax or payment shall be based on the following schedule:

If the energy efficiency is:		The tax or (payment) is:
Not over 5		$320
over 5 but not over 6		300
6	7	280
7	8	260
8	9	240
9	10	220
10	11	200
11	12	180
12	13	160
13	14	140
14	15	120
15	16	100
16	17	80
17	18	60
18	19	40
19	20	20
20	21	0
21	22	(20)
22	23	(40)
23	24	(60)
24	25	(80)
25	26	(100)
26	27	(120)
27	28	(140)
28	29	(160)
29	30	(180)
30	31	(200)
31	32	(220)
32	33	(240)
33	34	(260)
34	35	(280)
35 and over		(300)

(b) All ownership taxes collected pursuant to this section shall be deposited in an energy conservation reserve fund set aside and maintained by the state treasurer to be used for the prompt payment of moneys to persons purchasing energy efficient automobiles and of other expenditures as authorized in this section.

(c) Any payment to a new automobile owner pursuant to this section shall be made only from moneys available in the energy conservation reserve account. In the event such account is temporarily depleted, future payments shall be made when moneys are available in the order of receipt of the notice of entitlement to a payment, and any payments not made during the fiscal year shall carry over to the next fiscal period until paid. The energy conservation reserve account shall also be used by the department [of revenue] to defray the expense to counties for the administration of this section by reimbursement to them of a sum equal to twenty-five cents per paid new automobile registration as defined in this section. Any surplus moneys remaining in said reserve account after the making of payments to qualified automobile owners and reimbursements to counties shall be credited to the general fund. The general assembly shall make an annual appropriation to the department to pay all expenses and costs of the administration of this section by the department.

(5) In every odd-numbered year the department shall report to the governor and the general assembly concerning the revenue impact from the owership taxes and payments made pursuant to this section and the costs of administering such program.

(6) (a) In order to adequately inform the public of the provisions of this section:

(I) No dealer shall sell or offer to sell any new automobile which is subject to the provisions of this section unless the energy efficiency tax or payment applicable to such automobile is disclosed by the dealer prior to any such sale. Such disclosure shall appear in each contract, estimate, proposal, direct-mail statement, and in any other place which gives the purchase price or acquisition cost of such automobile.

(II) No dealer shall advertise or cause to be advertised through any communications medium any new automobile which is subject to the provisions of this section if such advertisement states the purchase price or acquisition cost of such automobile without also disclosing clearly and conspicuously the energy efficiency tax or payment applicable to such automobile.

(b) The provisions of this subsection (6) shall not apply to any contract, estimate, proposal, direct-mail statement, advertisement, or other disclosure involved or appearing in items of interstate commerce.

(c) The provisions of this subsection (6) shall not apply to the federal automobile labeling requirements of 15 U.S.C. 2006.

(7) The energy efficiency of each new automobile which is subject to the provisions of this section shall be contained in the manufacturer's certificate of origin or in the bill of sale of the vehicle when it is presented to the department at the time of initial registration.

SECTION 2. *Safety clause.* The general assembly hereby finds, determines, and declares that this act is necessary for the immediate preservation of the public peace, health, and safety.

Appendix 4–3

PITTSBURGH PARKING TAX ORDINANCE

No. 30

AN ORDINANCE

Providing for the general revenue by imposing a tax of 20 per centum (20%) upon the consideration paid by the patrons of a non-residential parking place for each parking transaction, to be collected from the patron by the operator of each non-residential parking place; requiring a license; providing for the levy and collection of such tax; prescribing the requirements for return and records, conferring powers and duties upon the Treasurer, imposing penalties; and providing for the exclusion of certain operators from the provisions of Ordinance No. 704 approved December 31, 1969.

THE COUNCIL OF THE CITY OF PITTSBURGH, UNDER THE AUTHORITY OF ACT NO. 511 OF 1965, AND ITS AMENDMENTS, HEREBY ENACTS AS FOLLOWS:

SECTION 1. This ordinance shall be known and may be cited as the "Parking Tax Ordinance."

SECTION 2. *Definitions*: As used in this ordinance, unless the context indicates clearly a different meaning, the following words and phrases shall have the meanings set forth below:

(a) "City"—The City of Pittsburgh.

(b) "Patron"—Any natural person who drives a vehicle to, into or upon a non-residential parking place as hereinafter defined for the purpose of having such vehicle stored for any length of time. "Patron" shall also include any natural person who has a vehicle in his custody or control collected from him by another for the purpose of having it stored at a non-residential parking place.

(c) "Person"—Any natural person, partnership, unincorporated association or corporation, non-profit or otherwise. Whenever used in any provision prescribing a fine or a penalty, the word "Person" as applied to partnership, shall mean the partners thereof, as applied to unincorporated associations shall mean the members thereof, and as applied to corporations, shall mean the officers thereof.

(d) "Non-Residential Parking Place" or "Parking Place"—Any place within the City, whether wholly or partially enclosed or open, at which motor vehicles are parked or stored for any period of time, in return for a consideration not including (i) any parking area or garage to the extent that it is provided or leased to occupants of a residence on the same or other premises for use only in connection with, and as accessory to, the occupancy of such residence, and (ii) any parking area or garage operated exclusively by an owner or lessee of a hotel, an apartment hotel, tourist court or trailer park, to the extent that the parking area or garage is provided to guests or tenants of such hotel, tourist court or trailer park for no additional consideration.

As used herein, the term "residence" includes (i) any building designed and used for family living or sleeping purposes other than a hotel, apartment hotel, tourist court or trailer park, and (ii) any dwelling unit located in a hotel or apartment hotel. The terms "hotel," "apartment hotel," "tourist court," "trailer park," and "dwelling unit" are used herein as defined in the Zoning Ordinance, Ordinance No. 192, approved May 10, 1958, as amended.

(e) "Month"—A calendar month.

(f) "Operator"—Any person conducting the operation of a parking place or receiving the consideration for the parking or storage of motor vehicles at such parking places, including, without limiting the generality of the above, any governmental body, governmental subdivision, municipal corporation, public authority, non-profit corporations or any person operating as an agent of one of the above.

(g) "Transaction"—The transaction involved in the parking or storing of a motor vehicle at a non-residential parking place for a consideration.

(h) "Consideration"—Refers to the payment or compensation, of whatever nature, received by the operator from the patron, upon an express or implied contract or under a lease or otherwise, whether or not separately stated, and whether paid in cash or credited to an account, for each transaction involving the parking or storing of a motor vehicle by the patron. The consideration shall not include the tax imposed and collected under this ordinance.

(i) "Treasurer"—The Treasurer of the City of Pittsburgh.

SECTION 3. *Imposition of Tax*: A tax for general revenue purposes is hereby imposed upon each parking transaction by a patron of a non-residential parking place, at the rate of 20 per centum (20%) on the consideration for each such transaction during the period April 1, 1973 to December 31, 1973 and thereafter from year to year on a calendar year basis. No operator shall conduct such transactions without complying with all of the provisions of this ordinance and without collecting the tax imposed here and paying it over to the City.

SECTION 4. *Annual License*: No operator shall conduct the operation of a non-residential parking place without obtaining for each parking place an annual license from the Director of the Department of

Public Safety of the City of Pittsburgh as required by Ordinance No. 15, approved February 1, 1971, within the time specified. Any operator not possessing such license for each parking place for the year 1973 or any following year shall obtain such license within thirty (30) days after the effective date of this Ordinance, and any person who intends to begin conducting the operation of a non-residential parking place shall obtain such a license before beginning such operation. At each parking place, the operator shall display the license in a conspicuous location at all times. Such licenses shall not be transferable between one operator and another or between one parking place and another. Any operator who ceases to conduct the operation of a parking place shall notify the Treasurer and return the license applicable therein.

SECTION 5. *Records*: Each operator shall maintain, separately with respect to each parking place, complete and accurate records of all transactions, of the total amount of consideration received from all transactions, and the total amount of tax collected on the basis of such consideration. Each operator shall issue to the person paying the consideration such written evidence of the transactions as the Treasurer may prescribe by regulations. Where consideration in a transaction is not separately stated, the operator shall maintain evidence and records necessary to segregate the consideration applicable to the transaction for the benefit of the patron and the Treasurer so that the proper amount of tax can be collected. Each operator shall afford the Treasurer and his designated employees and agents access to all such records and evidence at all reasonable times and shall provide verification of the same as the Treasurer may require.

The Treasurer and his agents are hereby authorized to examine the books, papers and records of any operator or probable operator in order to verify the accuracy of any return made, or, if no return has been made, to estimate the tax due. Every such operator, or probable operator, is hereby directed and required to give to the Treasurer, or any agent designated by him, the means, facilities and opportunity for such examination and investigations as are hereby authorized.

SECTION 6. *Return and Payments*: Each operator, on forms prescribed by the Treasurer, shall file by the 15th day of each month, returns for the preceeding month showing the consideration received with respect to each parking place during the preceeding month together with the amount of tax collected thereupon. At the time of filing the return, the operator shall pay to the Treasurer all tax due and collected for the period to which the return applies. Each operator shall collect the tax imposed by this ordinance and shall be liable to the City of Pittsburgh as agents thereof for the payment of the same to the City Treasurer.

SECTION 7. *Treasurer's Powers and Duties*: The Treasurer, on behalf of the City, shall receive and collect the taxes, fines and penalties imposed by this ordinance, and shall have the power, in the event that any operator has, in the judgment of the Treasurer, failed to pay over the amount of tax due, to collect the tax directly from the patron and charge the cost of collection to the operator, and shall maintain records showing the amounts received and the dates such amounts were received. The Treasurer shall adopt and enforce regulations relating to any matter pertaining to the administration of this ordinance, but not limited to, requirements for evidence and records and forms for application, licenses and returns.

SECTION 8. *Collection*: The Treasurer shall collect, by suit or otherwise, all taxes, interests, costs, fines and penalties due under this ordinance and unpaid. If the operator neglects, refuses, or fails to file any report or make any payment as herein required, an additional 10 per centum (10%) of the amount of the tax shall be added by the Treasurer and collected as a penalty. All taxes due and unpaid, shall bear interest at the rate of ___ % per month or fraction thereof from the date, they are due and payable until such time as they are paid.

SECTION 9. *Violation*: Without limiting the power of the City to prosecute any person violating any provision of this ordinance under the penal code of the Commonwealth, any person who violates any provision of this ordinance or any regulation adopted pursuant to it shall, upon conviction thereof before any alderman or magistrate, be liable for a fine of not more than Three Hundred Dollars ($300.00), or, in default of payment of such fine, shall be imprisoned in the Allegheny County Jail or Allegheny County Workhouse for a period not exceeding thirty (30) days. Each day's violation shall constitute a separate offense.

SECTION 10. The collection and transmittal of taxes imposed under this ordinance shall exclude the operator from all of the provisions of Ordinance No. 704 of 1969.

SECTION 11. If a final decision of a court of competent jurisdiction holds any provision of this ordinance, or the application of any provision to any circumstances to be illegal or unconstitutional, the other provisions of this ordinance or the application of such provision to other circumstances, shall remain in full force and effect. The intention of Council is that the provisions of this ordinance shall be severable and that this ordinance would have been adopted if any such illegal or unconstitutional provisions had not been included.

SECTION 12. *Effective Date*: This Ordinance shall take effect, April 1, 1973.

※ *Chapter 5*

The Cost of Energy and
the Poor and the Elderly

The fear that the poor and the elderly will suffer severe hardship if economic disincentives are enacted is one of the major obstacles to their adoption. Hard pressed by inflation, which the rising cost of energy has fueled, the disadvantaged already have trouble making ends meet. During recent winters, many poor and elderly persons have had to choose between eating and staying warm, and some have even frozen to death after their heat was shut off when they could not afford to pay their bills.[1] Given these circumstances, why should states and localities even consider proposals that, by deliberately raising the cost of energy, would only aggravate matters?

Imposing economic disincentives to encourage energy conservation will not necessarily and unavoidably worsen the plight of the poor and the elderly. Government has the means to alleviate any undue burden that disincentives may cause. Chapter 4, for instance, discusses several suggestions for using the revenues from automobile disincentives to mitigate their impact on the poor. Most of the suggestions—using the money to provide, improve, and expand transportation alternatives to the automobile or to reduce regressive sales and/or property taxes—would benefit not only the poor and the elderly but also middle and upper income persons. Other measures that could relieve the pressure of disincentives on the disadvantaged include tax rebates, such as those proposed in connection with the National Energy Plan's crude oil equalization tax and standby gasoline tax, and increases in the benefits provided by various public assistance programs.

Despite the reassurance that they need not suffer if economic disincentives are imposed, the poor and the elderly will probably continue to be wary of disincentives, particularly if they would increase bills for essential residential energy services such as electricity, natural gas, or home heating oil. This book does not propose any economic disincentives that would raise the cost of residential energy services. The decision not to include disincentives of this type arose partly from a belief that utility rate reform offers a better avenue for encouraging energy conservation by raising electricity and natural gas prices to replacement cost levels. (Rate reform issues are discussed in another book in this series, *Utility Pricing and Planning— An Economic Analysis*, by Frederick J. Wells.)

The decision also arose from a belief that disincentives are not needed as urgently in the residential sector as they are in other areas (e.g., the automotive segment of the transportation sector). Significant conservation efforts have been undertaken in the residential sector because, unlike gasoline prices, the prices of residential energy services have continued rising sharply since the embargo. The higher prices have, in fact, performed the same functions that disincentives would. The escalating prices have also revealed a serious problem— the hardship that the poor and the elderly experience paying the higher bills for residential energy services. The imposition of disincentives would probably intensify this problem.

This chapter examines three measures that states can employ to ease the burden of rising residential energy bills for the poor and the elderly—lifeline utility rates, energy stamps, and cash assistance programs. These measures could be used now to relieve the strain of rising prices, or they could be implemented concurrently with any residential disincentives enacted in the future. In the first instance, they should be regarded only as short term, stopgap measures. If adopted as a permanent solution to the problem, any of these measures, as energy prices continue to escalate, will exact an increasing toll on those who are financing the relief (i.e., other utility customers and taxpayers). Moreover, by themselves, these programs will do little to encourage, and could actually impede, progress toward the ultimate solution—reducing residential energy needs.

Conservation, achieved by having better insulated homes that are equipped with efficient appliances, is the only permanent way to alleviate the impact of higher electricity, natural gas, and heating oil bills on the poor and the elderly. Numerous federal, state, and local programs provide loans, grants, and income tax credits to underwrite energy-saving home improvements.[2] Unfortunately, even with these incentives, the process of upgrading the energy efficiency of the homes of the poor and the elderly will take several years to complete substantially. In the interim, many poor and elderly persons will suffer unless government somehow relieves the burden of rising energy prices.

RESIDENTIAL ENERGY PRICES
SINCE 1973

Prices of residential energy services have increased substantially since the embargo; the actual numbers reveal just how staggering these increases have been. In the years before the embargo, the prices paid for residential electricity, natural gas, and heating oil remained relatively stable, increasing at the same pace as other consumer prices.[3] From 1971 to 1972, for example, the Consumer Price Index (CPI) for fuels and utilities rose 3.8 percent, which paralleled the overall CPI increase of 3.5 percent. The embargo, however, destroyed this price stability and ushered in an era of sharply higher bills for residential energy services. From 1973 to 1974, the CPI for fuels and utilities soared 21.5 percent against an overall CPI increase of 10.5 percent. Home heating oil showed the largest increase, up 58.4 percent, and was followed by electricity, up 18.1 percent, and natural gas, up 12.5 percent.

After the embargo ended, the prices of residential energy services did not stabilize. Several factors combined to keep residential energy prices spiraling upward. The recession triggered by the embargo brought hard times for electric utility companies,[4] and they sought rate increases to bolster their declining fortunes. Although they had quadrupled their prices by the end of the embargo, the Organization of Petroleum Exporting Countries (OPEC) raised their prices again in late 1976. Harsh winter weather, particularly during the 1976–1977 winter, increased the nation's reliance on expensive imported oil and exacerbated natural gas shortages in the interstate market.[5] The combined effect of these events has been reflected in the higher electricity and natural gas prices consumers have paid since 1974.

In the four year period beginning in 1974, electric and natural gas rates rose a total of $48.3 billion.[6] In contrast, these rates rose a total of only $6 billion in the quarter century prior to 1974. Higher fuel costs have been primarily responsible for the higher rates. Under fuel adjustment clauses, utilities are allowed to pass on to consumers increases in fuel costs automatically without petitioning public service commissions for rate increases. Of the $48.3 billion increase in electricity and natural gas rates from 1974 through 1977, $35.6 billion was passed through under fuel adjustment clauses, and public utility commissions granted $12.7 billion in rate hikes after formal proceedings in rate cases. Electric utilities received $23.7 billion in fuel adjustment increases, and gas utilities received $11.9 billion under these clauses.

Two events that occurred in late 1978 insured that the upward spiral of gas and electric rates would continue in the foreseeable future. First, in October 1978, Congress passed the Natural Gas Policy Act of 1978 as part of the National Energy Act.[7] This Act allows prices of newly discovered gas to rise at the rate of inflation plus 3.5 or 4 percent a year until 1985. Assuming 6 percent inflation, this formula would allow the price of gas to increase almost 90 percent between October 1, 1978, and January 1, 1985.[8] Second, in December 1978, OPEC announced a 14.5 percent increase in their prices for 1979. Home heating oil consumers and customers of electric utilities that burn oil to generate a major portion of their power will be hit hardest by the OPEC action.

Although residential energy costs have increased and will increase for all consumers, the rising prices have had and will have a greater adverse effect on low and fixed income households than on more affluent households. Studies have shown that although poor households consume less electricity and natural gas than other households, they spend a higher proportion of their disposable incomes to fulfill their energy requirements. One study conducted by the Department of Health, Education, and Welfare, in cooperation with the Federal Energy Administration (FEA), found that a low income family used 11 percent of its income to purchase electricity and natural gas, but that a family with a $16,000 income spent only 2 percent of its income on those items.[9] Table 5-1 illustrates the regressive effect of rising electricity rates. The certainty that residential energy costs will continue to escalate emphasizes the need for government action to prevent these costs from increasing the burden they impose on the poor and the elderly.

LIFELINE RATES

Under the traditional declining block rate structures, electric utility customers generally pay a decreasing price per kilowatt hour (kwh) for each succeeding "block" of kilowatt hours used. With this type of rate structure, low use customers, often residential customers, pay more per kwh for the electricity they actually consume than high use customers. In 1974, for example, residential electricity consumers paid an average price of 3.04 cents per kwh, as compared to commercial consumers who paid 2.99 cents per kwh and industrial consum-

Table 5-1. Average Annual Electricity Expenditures by Disposable Household Income, 1973 to 1975.

	1973	*1974*	*1975*
Less than $3,400	$149	$172	$183
$ 3,400- 6,899	163	183	200
$ 6,900-10,499	186	215	228
$10,500-15,199	207	240	254
Expenditures as a Percent of Income			
Less than $3,400	8.8	10.1	10.8
$ 3,400- 6,899	3.2	3.7	3.9
$ 6,900-10,499	2.1	2.5	2.6
$10,500-15,199	1.6	1.9	2.0

Note: Households with disposable income $15,200 or greater are omitted from the table. Percentages are calculated on the basis of the midpoint of the income classes.

Source: Federal Energy Administration (FEA), Household Energy Expenditure Model; and S. Mintz, "An Explanation of Electric Utility Finance and Its Effects on the Residential Consumer" (paper presented at the 22nd Annual American Council on Consumer Interests Conference, Atlanta, Georgia, April 8, 1976), Table VII at 23.

ers who paid 1.64 cents per kwh.[10] This disparity in rates has led some advocates of electricity rate reform to argue that low use customers, including the poor, subsidize the electricity consumption of wealthier, high use customers. The dramatic increase in electricity costs since 1973 has magnified the inequity of the subsidy, and lifeline rates have been offered as a means of redressing the balance.

Lifeline rates would provide residential customers an initial block of kwh (or therms of natural gas) at a low unit cost. The initial ("lifeline") block would represent an estimate of what quantity of energy would be sufficient to satisfy subsistence energy needs. Subsistence energy requirements are generally defined as the minimum energy needs of an average residential customer for space and water heating, lighting, cooking, and food refrigeration.[11] Table 5-2 compares typical declining block and lifeline electric rate structures. This section will examine lifeline electricity rate proposals to determine whether they would achieve the objective of relieving the burden of high electricity prices on the poor and elderly.

Several mechanisms exist for introducing lifeline rate structures. Existing rates for the lifeline block (whatever that is determined to be) can be lowered, or they can be frozen while the rates for other levels of consumption are increased.[12] In the first case, utilities will

Table 5–2. Declining Block Rates and Lifeline Rates.

Declining Block Structure	
Rate	*Consumption Level*
$2.25	first 25 kwh
3.50¢ per kwh	next 75 kwh
3.25¢ per kwh	next 100 kwh
3.00¢ per kwh	next 200 kwh
2.50¢ per kwh	next 400 kwh
1.50¢ per kwh	all additional kwh
Lifeline Structure (California)	
Rate	*Consumption Level*
$1.50	Customer Charge
2.50¢ per kwh	first 240 kwh
4.20¢ per kwh	next 60 kwh
4.60¢ per kwh	over 300 kwh

lose revenues when rates are lowered to provide the lifeline benefits. This revenue loss will probably lower utilities' actual rate of return below their allowed rate of return. To recover their lost revenues, utilities will probably be permitted to increase rates for residential customers on consumption levels above the lifeline block and/or for commercial and industrial customers. If the second approach for instituting lifeline is chosen, the rate hikes granted for levels of consumption above the frozen lifeline block will probably be higher than they would be if rates for the lifeline block were also increased. The higher rates that someone bears as a result of the adoption of lifeline rates constitute the lifeline burden.

Most utilities have vigorously opposed lifeline proposals because they do not feel it is their function to remedy the nation's social problems.[13] Disagreeing with the utilities, federal and state legislators and regulatory officials have seriously considered, and in a few cases adopted, lifeline proposals. At the federal level, the ninety-fourth Congress considered several bills to reform electric utility rates.[14] Some bills would have mandated that regulatory commissions adopt lifeline rates while others would have required only that the commissions consider lifeline rate structures. During debate on President Carter's National Energy Plan in the ninety-fifth Congress, the Senate passed a provision requiring utilities to establish lifeline rates for the elderly. The House–Senate energy conferees, however, dropped this requirement for one that directs state utility commissions to consider lifeline rates.[15]

Since the escalation in electricity and natural gas prices began in 1974, lifeline proposals have been considered in many states. In 1975, twenty-eight bills to mandate implementation of lifeline rates were considered in seventeen states, and in 1976 at least thirty-two such bills were pending in eighteen states.[16] Lifeline proposals contined to receive considerable legislative attention in 1977 and 1978. In addition to action on the legislative front, numerous state public utility commissions have studied lifeline proposals either on their own initiative or in response to consumer participation in rate proceedings.[17] In other states, citizen groups, stymied by legislative or administrative hostility or indifference to lifeline proposals, placed ballot initiatives concerning lifeline or fair share rates before the voters in 1976 and 1978.[18]

Some states have advanced beyond the mere consideration or study stage and have adopted lifeline rates. During 1976, Maine had a one year demonstration project involving lifeline rates for low income senior citizens.[19] As a result of the Miller-Warren Energy Lifeline Act of 1976, California now has lifeline rates for both electricity and natural gas.[20] In 1978, New Jersey passed legislation extending lifeline rates for electricity and gas to customers meeting an income eligibility test.[21] Besides these states, at least two localities, Yellow Springs, Ohio, and Aztec, New Mexico, have lifeline rates for customers of their municipally owned utilities.[22] In several states, public service commissions have taken the initiative in ordering utilities to adopt lifeline rates.[23]

The principal reason for the interest in and adoption of the lifeline rate concept has been its purported efficacy as a means of alleviating the burden that rising electricity and gas prices place on the poor and the elderly. Two collateral advantages of lifeline rates, according to lifeline advocates, are their administrative simplicity and their potential for encouraging energy conservation.[24] Several analyses of lifeline proposals, however, raise considerable doubt about whether lifeline rates really possess this trio of virtues.

As a strategy for relieving the burden of high utility prices on the poor and the elderly on fixed incomes, lifeline rates are based on the assumption of a close correlation between a utility customer's income and his or her utility consumption. The assumption is that low income customers are low use customers (and vice versa) and that reducing utility rates for the latter will ease the burden of rising rates on the former. Although some studies have suggested that income and utility use are closely related, several have demonstrated that family size, dwelling conditions, and climate are more important determinants of electricity consumption.[25]

Seizing upon the questionable validity of the key assumption underlying the lifeline concept, lifeline critics have claimed that the basic lifeline proposals (i.e., those that would freeze or lower rates on the first several hundred kwh of consumption for *all* residential consumers) would provide inadequate coverage while bestowing unwarranted benefits. For example, electric water heaters, which typically consume about 300 kwh per month, would devour the entire lifeline block under many proposals. In twenty-three states, mainly in the South, Midwest, and Northwest, at least 25 percent of the poor (income under $4,000) possess electric water heaters.[26] Electric space heat is less widespread among the poor. In only five states—Florida, Tennessee, Nevada, Oregon, and California—do more than 20 percent of the poor live in electrically heated homes.[27] Many small farmers, who use large amounts of electricity, have not grown wealthy as a result of the higher food prices of recent years. In eleven states, at least 5 percent of the poor are farmers.[28] Farmers who pay residential class rates and poor households with electric heat and/or water heaters will probably not receive the benefits of lifeline rates and may even bear some of the lifeline burden.

Not only does the assumed correlation between income and energy use mean that lifeline rates will not reach a significant portion of poor households in some states, it also means that some affluent households may reap conservation-discouraging benefits. For example, a wealthy couple living in an apartment in which they pay their own electricity bills may consume less electricity than a low income family living in a poorly insulated single family home. The latter may not receive any lifeline benefits, and the former, who can easily absorb the burden of higher electricity bills, may respond to reduced monthly bills by increasing their consumption. According to one study, as many as 20 percent of the households with at least upper middle incomes consume less than the average amount of electricity.[29]

This problem of rate structures that benefit the affluent as well as the poor could be solved by adopting an income-related eligibility standard in states without prohibitions against charging differing rates for the same service. For instance, eligibility for lifeline rates could be restricted to food stamp or public assistance recipients. Basing eligibility for lifeline rates on actual participation in one of these welfare programs would eliminate the need for income verification, one administrative difficulty of income-related lifeline proposals. Offsetting the gain in administrative simplicity, however, is the injustice of denying lifeline benefits to those persons who qualify for, but do not register to receive, food stamps or public assistance. Of course, any lifeline rate structure with an income-related eligibility standard would not be as easy to administer as one without any standard, but the equity that the standard provides may outweigh the added administrative inconvenience.

Another instance of the inadequate coverage provided under the basic lifeline proposals does not arise out of the inexact correlation between income and electricity use. Most of the lifeline rate proposals will not benefit the millions of poor renters whose apartment buildings do not have separate electricity meters for each apartment. These persons pay their utility bills indirectly as a part of their rent because their building's master meter only records the building's total monthly electricity consumption. In twenty-seven states (including nine of the ten most populous), at least 10 percent of the poor are tenants whose electricity costs are included in apartment rents.[30] In addition to not receiving the benefits of lifeline rates, these renters will probably bear the burden of lifeline rates if their landlords are billed at the higher bulk residential or commercial rates that are imposed to recover the revenue lost through lifeline reductions. Limiting revenue recovery to the residential class would eliminate this potential problem if apartment buildings are billed at commercial class rates.

Calculating electric bills for apartment buildings with master meters in a manner that takes lifeline benefits into account would eliminate the hardship that lifeline rates could cause for poor renters.[31] First, a building's monthly electricity consumption could be divided by the number of apartments in the building to determine the average consumption per apartment. Next, using this average consumption figure, a bill for that level of consumption, including any lifeline benefits, could be calculated. Finally, this bill could be multiplied by the total number of apartments in the building to determine the landlord's electricity bill.

Admittedly, this suggestion for extending lifeline benefits to poor renters in master-metered apartment buildings has its shortcomings. First, it would lower electricity bills of buildings with affluent tenants as well as those housing poor renters. Second, it would not guarantee, by itself, that rents would be held steady, much less lowered, if this method of calculation reduced a building's total electric bill. Nevertheless, until the practice of master metering is eliminated,[32] adopting this approach for billing master-metered apartments would at least remove one source of pressure on landlords to raise rents in order to recoup increasing costs.

Making the necessary adjustments to correct the flaws in the basic lifeline proposals will sacrifice one of the asserted advantages of lifeline rates, their administrative simplicity. The more complex the lifeline rate structure becomes, the more it has in common, from an administrative viewpoint, with other programs for assisting the poor and the elderly to pay their utility bills.

According to lifeline advocates, a third advantage of lifeline rates is their potential for encouraging energy conservation. By lowering rates for a subsistence quantity of electricity while raising rates for additional units, an incentive to conserve electricity is created. One FEA study estimated that national implementation of lifeline rates would produce annual residential savings of approximately thirty million barrels of crude oil and total annual savings, including commercial and industrial savings, of over one hundred million barrels.[33]

Lifeline critics, however, have disputed assertions concerning the concept's energy conservation potential. Analysis of many early lifeline proposals revealed that the lifeline rate structure might actually encourage increased consumption. One study of the lifeline rates proposed for Vermont concluded that any residential customer using less than 1,400 kwh, which was twice the statewide average monthly residential consumption, would pay a lower electric bill under the proposed lifeline structure than under the prevailing rate structure.[34] This consumption-stimulating result occurred because the rates for levels of consumption above the lifeline block did not increase rapidly enough. This problem has a simple solution.

Many of the basic lifeline proposals increase rates for commercial and industrial users to recover some of the revenues lost in the reduction of rates for the lifeline block. If, instead, a lifeline proposal requires the residential class alone to make up the lost revenues, the Vermont result could be avoided. By concentrating the rate increase necessary to recover the lost revenues on the range of consumption between the end of the lifeline block and a point of consumption below which 75 percent of all residential consumers fall, a rate structure with a powerful energy conservation incentive could be designed. Rates for commercial and industrial customers could be raised independently of the lifeline rate revision to tap the substantial potential for conservation that exists among these users.[35]

An alternative approach to the objective of instilling an energy conservation element in lifeline rate structures would restrict eligibility for the lower lifeline rates to those customers who consume less than the lifeline ceiling. For example, if the lifeline block were 500 kwh, households that used less than 500 kwh a month would be billed at the low lifeline rate. Households that used more than 500 kwh would pay a higher rate for the entire amount consumed. Structuring lifeline rates in this manner would encourage energy conservation, but would not accomplish the goal of easing the burden of high energy costs for the poor. According to one study, if this alternative (with a lifeline block of 500 kwh) were implemented nationwide, over 25 percent of the poor in thirty-nine states would not qualify for the low lifeline rate.[36]

As this discussion of lifeline rates has attempted to demonstrate, most of the shortcomings of basic lifeline proposals pointed out by lifeline critics can be remedied by modifying the rate structure design. The redesigned lifeline rate structure—income-related eligibility standards with adjustments for poor tenants, poor farmers, and poor families with electric space and/or water heaters—may create a serious administrative quagmire. Even with these modifications, lifeline rates provide no relief to the poor and the elderly who must rely on other fuels for heating and cooking. Energy stamps and cash assistance programs, although perhaps more difficult to administer, can provide not only broader coverage, but also a higher level of benefits. Because of these alternatives, lifeline rates are not the best means of alleviating the burden of higher residential energy costs on the poor and the elderly.

Lifeline rates may, however, have a role to play as part of a comprehensive reform of electric utility rate structures. In August 1978, the New York State Public Service Commission issued an opinion concerning lifeline rates.[37] The commission flatly rejected proposals for lifeline rates grounded on equity considerations.[38] The commission, though, showed considerable interest in cost-justified lifeline proposals. The proposal that intrigued the commission would require utilities to charge customers rates based on the marginal cost of providing the service. The proposal assumed that marginal cost pricing would generate revenues in excess of those necessary to enable utilities to earn their allowed rate of return. To remedy the problem of surplus revenues, the proposal suggested that the revenues be returned to consumers in a way, such as through a lump sum rebate, that would not alter the marginal cost price signal. Although the commission refused to embrace "as a matter of principle" the rebate proposal, it ordered an experiment to test its potential effectiveness. The commission concluded, however, that in any event, the rebate proposal would not be the best solution to the problem of high energy costs for low income persons and that a legislative solution would offer the optimum prospect for immediate and effective relief.

ENERGY STAMPS

Energy stamp programs, similar in design to the federal food stamp program, would provide energy stamps or vouchers to low income households to enable them to purchase residential energy services sufficient to satisfy their basic energy needs.

In the aftermath of the embargo and during the severe winter of 1976–1977, federal energy stamp proposals were extensively studied, and legislation to create a national energy stamp program was introduced in Congress.[39] During the 1974–1975 winter, federally subsidized energy stamp demonstration projects were undertaken in Denver, Colorado, and in Pennsylvania's Lehigh Valley. The Lehigh Valley program, Project HELP (Heating for the Elderly and Low-Income Persons), was funded with a grant from the federal Community Services Administration (CSA) and administered by the Community Action Committee of the Lehigh Valley, Inc.[40] To offset the increase in heating costs that had occurred since the 1973–1974 winter, this program provided a $50 subsidy to eligible households (i.e., individuals or families at or below the Office of Economic Opportunity poverty guidelines) who paid directly for their home heating fuel or electricity. Participants received the subsidy by paying $25 for a booklet of energy vouchers worth $75. Almost $250,000 was distributed to eligible families, and the administrators of the program estmated that 65 percent of the eligible households in the area participated.

In 1976 the Lorain County, Ohio, Community Action Agency established an energy stamp program with a $7,500 trial grant from the CSA.[41] Similar in design to Project HELP, the program provided assistance to low income families with actual natural gas emergencies. To qualify for the stamps, applicants not only had to fall within the federal poverty standard, but also had to demonstrate that an actual emergency (e.g., receipt of a termination of service notice from the gas company) existed. The stamps, which cost $25 for a booklet of fifteen worth $75, could be used to pay gas (but not electric) bills.

At the state level, little action has been taken toward establishing state-financed and administered energy stamp programs. Vermont legislators showed no enthusiasm for a bill introduced in 1975 to authorize the human services agency to establish and administer an energy coupon program.[42] During 1976, the Michigan legislature considered a bill to create a utility services stamp program for senior citizens.[43] The Michigan Department of Commerce and Public Service Commission recommended that the bill be amended to extend coverage to all poor households and to permit recipients to use the vouchers to pay for home heating fuels as well as for utility services.[44] With these amendments, the bill, which is similar to a proposal made by William S. Rosenberg, a former chairman of the Michigan Public Service Commission (referred to hereafter as the "Rosenberg proposal"),[45] provides a model for states contemplating an energy stamp program. (See the appendix to this chapter for the text of the amended bill.)

Even in the absence of legislative interest in energy stamps, many utility companies have touted the stamp programs as a better answer than rate reform to the problem of alleviating the adverse impact of rising energy prices on the poor. For example, the Florida Power Corporation developed an energy stamp proposal in 1975.[46] Under this proposal, all families qualified for assistance under the Aid to Families with Dependent Children and food stamp programs would receive a monthly voucher. Utilities would redeem these vouchers for the actual amount of the monthly electric bill up to a maximum charge equivalent to 600 kwh consumption. The state would reimburse the utilities for the redemption value of the vouchers. Florida Power presented this proposal to the state legislature, which held hearings but never drafted any legislation to implement the program.

Although utility companies have enthusiastically advocated energy stamps because of their belief that the burden of social relief should be borne by taxpayers and governments, an energy stamp program would, in any event, avoid several of the problems connected with lifeline rates. Energy stamps, linked directly to personal income, focus their benefits on the needy and do not subsidize the energy expenditures of the rich. Energy stamps can aid poor renters who, even though they do not pay utility or heating bills directly, could use the stamps to defray the utility and heating fuel costs built into their rents. Moreover, energy stamps can provide a more meaningful level of assistance than lifeline rates that would reduce electric bills by only $2 to $8 a month.[47] Finally, energy stamps would not deny benefits to poor households that use home heating oil rather than electric or natural gas heat because the stamps could be used to buy any type of residential energy.

Although energy stamp programs have significant advantages, they also have some serious disadvantages. An energy stamp program patterned on the food stamp program would probably be costly to administer and subject to many of the problems that have plagued the food stamp program. Even if steps are taken to reduce administrative costs (e.g., adopting food stamp eligibility standards and delegating administration of the program to an existing agency such as the department of social services), an energy stamp program would add another layer to the state bureaucracy and increase the cost of state government.

The major obstacle to the enactment of an energy stamp program, however, is not the cost of administering the program, but the cost of the benefits themselves. Economic disincentives focused on residential energy use would be one logical source of funds for an energy stamp program, but few, if any, states are likely to be considering this type of disincentive. Many states, though, currently impose a sales or use tax on residential energy sales. Revenues from these taxes could be devoted to financing energy stamps. A third alternative would involve using general tax revenues. Whatever choice is made, proponents of energy stamps should keep in mind that, in the wake of Proposition 13, any proposals to raise taxes to increase welfare benefits can expect a cool reception.

Assuming that the political problems of energy stamps can be resolved, any state interested in creating an energy stamp program should consider the following basic issues in designing the program and drafting enabling legislation.

Scope of Program

A state legislature must decide what types of energy services participants will be able to purchase with the stamps. The fairest approach would allow stamp recipients to buy any kind of energy that could be used in the home. Thus, the stamps could be used to purchase coal, kerosene, and wood as well as electricity or gas.

The legislature must also decide whether to issue stamps not only to persons who pay for their fuel and utilities directly but also to renters who pay their utility and heating costs as part of their rent. Including renters and allowing them to apply the stamps toward their rent is the more equitable alternative, but unscrupulous landlords may be tempted to reap a windfall by raising their rents in the amount of the stamp payments. Prohibiting this practice and backing up the prohibition with penalties such as substantial fines or even imprisonment may remove the temptation to steal the benefits provided by the stamps.

Eligibility Criteria and Certification Procedures

In establishing eligibility criteria for an energy stamp program, the legislature can reduce a state's administrative costs by selecting an existing set of standards, such as those for the federal food stamp or state public assistance programs. (The criteria selected may influence the decision, discussed below, concerning whether the amount of benefits should vary according to family size and income or should be distributed to all participants equally.) Provisions for verifying an applicant's eligibility for certification to participate in the program are also necessary and should include administrative due process hearings for applicants who are denied certification or who have their certification revoked.

Amount of Subsidy

In determining the size of the individual subsidy, the legislature should strive to accomplish the goal of relieving the poor of the staggering burdens of rising energy prices without creating a remedy that discourages energy conservation. Reimbursing the poor for the total increase in their energy costs would remove any effect that the higher prices might have as a disincentive. This consideration aside, however, total reimbursement would impose an extremely heavy burden on a state's treasury. Michigan estimated that subsidizing the entire increase in home energy costs of its residents eligible for the food stamp program would cost $150 million annually. A sliding scale of benefits, however, calculated on the basis of the increase in the subsistence residential energy budget, would cost Michigan only $29.6 million and would still provide a subsidy sufficient to enable the poor to purchase basic residential energy services.[48]

Another fundamental question related to the amount of the subsidy is whether each participant should receive an equal amount of stamps or whether the subsidy should vary according to family size and income. Project HELP took the first approach, and the Rosenberg proposal recommended the second. Although issuing an equal amount of stamps to all eligible participants would contribute to administrative simplicity, a sliding scale of benefits would be more equitable. The Rosenberg proposal developed its sliding scale after establishing the maximum subsidy, which was an amount equal to the weighted average of price increases in all residential energy forms. To calculate the weighted average, the Rosenberg proposal first determined the cost increase per household for subsistence amounts of each of the five types of energy used for space heating and for the subsistence level of electricity necessary for nonheating uses. Next, weighted cost increases for each of these six items were calculated and added together to arrive at the overall weighted average–maximum subsidy of $171. Having determined the maximum subsidy, the Rosenberg proposal developed a subsidy schedule in which the amount of the subsidy increases with the family size but decreases as family income rises (see Table 5–3). The Rosenberg program produces more equitable results than the Project HELP program because it takes into account not only income, but also the availability and

Table 5–3.   Energy Stamp Subsidy Schedule.

Family Size	$ 0– 1,000	$1,000– 2,000	$2,000– 3,000	$3,000– 4,000	$4,000– 5,000	$5,000– 6,000
1	$ 99	$ 89	$ 0	$ 0	$ 0	$ 0
2	107	97	87	0	0	0
3	115	105	95	85	0	0
4	123	113	103	93	83	0
5	131	121	111	101	91	81
6	139	129	119	109	99	89
7	147	137	127	117	107	97
8	155	145	135	125	115	105
9	163	158	145	133	123	113
10	171	161	151	141	131	121

Total Cost of Program $29,645,701

Program Population 304,712 Families

Source: William G. Rosenberg, Chairman, Michigan Public Service Commission, "Impact of Higher Home Energy Costs on Low Income Families and Alternative Public Programs to Reduce Hardships" Michigan Public Service Commission, East Lansing, Michigan (June 25, 1975) at 24.

differing costs of alternative energy sources and the effect of family size on residential energy consumption.

Operation of Program

Having determined the scope of the program, its eligibility requirements, and the amount of the subsidy, the legislature should develop, or authorize an appropriate administrative agency such as the department of social services to develop, guidelines for the program's operation. Among the questions to be resolved are:

- Who issues the stamps—e.g., does the administrative agency control distribution or delegate issuing responsibility to private outlets such as banks and credit unions?

- How the stamps are issued—e.g., must recipients pay a fee for the stamps (for example, a $25 fee for $75 worth of stamps) or do they simply receive the net benefits ($50 worth of stamps), thus eliminating the problem of participants not having sufficient cash to cover the fee?

- How often are the stamps distributed—i.e., are they issued periodically or is the entire annual allocation distributed at once, perhaps at the beginning of winter?

- How many recipients use the stamps—i.e., may they spend the total benefits to which they are entitled to pay one month's bill or are they limited to spending a certain percentage of their benefits each month?

- How do fuel dealers, utilities, and landlords who receive the stamps redeem them—i.e., do the payees present the stamps to the administrative agency for reimbursement or can they simply deposit the stamps, like cash or checks, in their bank accounts?

One of the most serious problems that could arise from an energy stamp program is the development of a black market for the stamps. One way to combat the fraudulent use of the stamps is to design stamps similar to travelers checks, the approach used by Project HELP. Issuing stamps in the name of one member of the household and requiring a countersignature would reduce the stamps' value to thieves and other unauthorized users. Using special currency type paper would make counterfeiting more difficult. Refusing to issue duplicates for allegedly lost or stolen stamps would eliminate one means of fraudulently obtaining additional stamps for use or resale, but it would impose a hardship on recipients with legitimate losses. All of these precautions would be taken in anticipation of the development of an illegal black market for the stamps.

Designing the program to include a legal white market for the sale of stamps might also reduce the potential profitability of a black market. A white market for energy stamps would allow the recipients of the stamps to convert them into cash at designated redemption centers, such as banks. Permitting the cashing of the stamps would make it easy for renters who do not pay their utilities separately to apply the value of the stamps to their rent. The sell back mechanism could also foster energy conservation by giving the stamp recipients the option of reducing their energy expenditures in order to spend the stamp benefits on other purchases. To eliminate some of the fraud that a white market would encourage, the program should require that individuals cashing the stamps present some form of identification verifying their entitlement to the stamps.

An energy stamp program complete with protective measures, such as travelers check type stamps and a white market mechanism, would entail substantial administrative cost and inconvenience. Rather than incur the bureaucratic hassles that would arise from having to distribute, collect, and redeem the stamps, some states have opted to skip the stamp stage and simply provide cash to the poor and the elderly to help them pay their residential energy bills.

CASH ASSISTANCE PROGRAMS

Since 1975 many states have considered and several states have established cash assistance programs for poor and/or elderly persons to enable them to pay increasing residential energy bills. This section compares programs enacted in Connecticut, Oregon, and Kentucky.[49]

Connecticut has been in the forefront in developing cash assistance programs. In early 1975, the Connecticut legislature considered, but never passed, a bill to authorize cash payments to eligible households to help them meet emergency energy needs.[50] Households that received, or were eligible to receive, food stamps and that spent more than 50 percent of their incomes for shelter costs, including utilities other than telephones, would have been eligible for the energy payments. Each household would have received $30 a month for the six winter months from November 1974 through April 1975, paid in a lump sum of $180 after the legislation passed. The bill also directed the welfare commissioner to obtain the maximum federal reimbursement for these energy payments.

If this bill had passed, it would have had an effect similar to that of a tax credit or rebate. The beneficiaries would have received the money long after the utility bills were due, the time when their need was most acute. A more serious shortcoming of the bill, however, was its failure to differentiate between those poor for whom the burden of higher energy prices was relatively light and those for whom the burden was heavy. An eligible family of seven with an annual income of $2,500 would have received the same payment as a family of three with an income of $3,500. In comparison, under the Rosenberg energy stamp proposal, the first family would receive benefits of $127 and the second $85.

Instead of this bill, Connecticut legislators passed a bill providing for direct payments to fuel dealers and utilities that supplied energy services to poor families.[51] First established for the fiscal year ending June 30, 1975, the program was extended through the fiscal year ending June 30, 1977.[52] The legislation empowered the state welfare commissioner to "provide for direct payments to venders for fuel and utility costs of eligible needy families with children who are beneficiaries of the aid to families with dependent children and the general assistance programs." The legislation instructed the commissioner to structure the payments so as to qualify the program for federal funds under the Emergency Assistance Program.[53]

Under the terms of the legislation, administrative regulations were to specify "(1) the eligibility conditions imposed for the receipt of emergency assistance; (2) the emergency needs for which payments may be made; and (3) the methods of providing such payments." The regulations[54] contain two eligibility criteria: (1) receipt of federal Aid to Families with Dependent Children (AFDC) or general assistance and (2) the existence of an emergency at the time the request for emergency assistance is made. An emergency is deemed to exist when one of the following three conditions is met: (1) utility bills have not been paid in full, and the utility has threatened a shut off of service; (2) service has actually been discontinued; or (3) the family finds it impossible to locate a company to deliver oil. Because any of "these situations are considered a serious hazard to the health of the children in the family, [i]mmediate action may be taken to pay the amount of the outstanding bills in excess of the allowance for fuel and/or utilities included in the [AFDC or general assistance] award each month for these items." The regulations also specify the verification procedures that are required before the caseworker can telephone the utility to inform the company that payment is forthcoming and that service should be restored or continued without interruption.

In 1978 Connecticut passed legislation to create an emergency fuel assistance program for low income families who do not receive AFDC or general assistance.[55] Only households with incomes no higher than 125 percent of the federal Community Services Administration (CSA) poverty guidelines for nonfarm households are eligible for the program. Each eligible family is limited to a maximum of $100 assistance.

In 1977, as part of its comprehensive energy conservation package, Oregon created a utility-heating fuel rate relief program for the low income elderly.[56] Tied in with the state's homeowner and renter property tax refund program, the rate relief program provides a $50 refund to taxpayers who (1) are sixty years of age or older, (2) have a household income of less than $5,000, and (3) are eligible and file a claim for a homeowner or renter property tax refund. For 1977–1979, $7 million are appropriated to fund the rate relief program.

Oregon has maximized administrative simplicity by linking the rate relief program to the property tax refund program. Giving every eligible taxpayer a flat $50 rate relief refund also contributes to administrative simplicity, but it may do so at the expense of equity. A sliding scale of benefits, based on variables such as income and family size, would probably be more equitable.

In 1978 Kentucky enacted legislation establishing "an energy cost assistance program for making money payments to or on behalf of [low income elderly persons] for the purpose of purchasing or supplementing the cost of energy for household use."[57] To be eligible to receive assistance under this program, a person must either be (1) sixty-two years of age or older and have an income equal to or less than 125 percent of the applicable CSA poverty guidelines or (2) blind or permanently and totally disabled and receiving supplemental security income benefits or eligible for medicaid benefits. Persons who meet these eligibility criteria are entitled to receive the payments (or have them paid on their behalf) regardless of whether they pay energy suppliers directly or pay their energy costs as part of their rents. The legislation directs the department for human resources to establish by regulation a schedule of payments and benefit levels based on income, resources, and family age. Payments will be made four months a year, presumably during the height of the winter. The legislation also authorizes the department "at its discretion and to ensure the proper application of the funds appropriated" to make the payments to energy vendors or suppliers directly or jointly to vendors and eligible recipients. For the 1978–1979 and 1979–1980 fiscal years, annual appropriations from the state's general fund are $5 million.

CONCLUSION

The Kentucky program incorporates many of the best features of the various proposals for helping the poor and elderly to pay for their residential energy needs. It avoids the potential administrative quagmire of energy stamps, but does not sacrifice equity (i.e., variable benefits) to preserve administrative simplicity. Unlike lifeline rates, it will not confer benefits on the wealthy, and it covers tenants whose energy costs are part of their rents as well as those persons who pay their residential energy bills directly. The implementation of the Kentucky program should be watched closely by other states because, with certain modifications, such as removal of the age limitation, this program could serve as a model for states trying to develop an effective means of alleviating the hardship for the poor and the elderly caused by rising residential energy costs.[58]

NOTES TO CHAPTER 5

1. In November 1978 a citizen-labor energy coalition that includes the National Council of Senior Citizens and the National Clients Council alleged that at least 200 persons died in recent harsh winters because of utility shutoffs. UPI, "Citizens Group Asks for State Bans Against Turning Off Heat in Homes," *Washington Post*, November 15, 1978, p. A–15.

2. For example, the federal Community Services Administration is authorized to establish programs to provide financial assistance to encourage "the weatherization of old or substandard dwellings, improved space conditioning and insulation." 42 U.S.C. § 2809(a)(12). The Energy Conservation and Production Act of 1976 (P. L. 94—385) established a $200 million grant program to assist low income persons to insulate their homes. 42 U.S.C. §§ 6861—6872. This program, which the U.S. Department of Energy administers, was extended for two years by part of the National Energy Act, the National Energy Conservation Policy Act of 1978 (P. L. 95—619). This Act also creates an insulation grant program under the Farmers Home Administration to assist low income families in rural areas, a grant program administered by the U.S. Department of Housing and Urban Development to improve the energy efficiency of multifamily public housing projects, and a program of loans to encourage the installation of solar energy equipment. Another part of the National Energy Act, the Energy Tax Act of 1978 (P. L. 95—618), authorizes tax credits for certain energy-saving home improvements.

Another book in this series, *Building to Save Energy: Legal and Regulatory Approaches*, by G.P. Thompson, discusses many of the efforts undertaken at the state and local level to help the poor make energy conservation improvements to their homes.

3. The Consumer Price Index data in this paragraph is from *A Study of the Effects of Rising Energy Prices on the Low and Moderate Income Elderly*, Final Summary Report, prepared for Consumer Affairs/Special Impact Office, Federal Energy Administration (Washington, D.C.: U.S. Government Printing Office, March 1975), pp. 2.38—2.40; and "The Impact of Rising Energy Prices on the Poor Between 1973—1974," an undated Federal Energy Administration release.

4. In the aftermath of the embargo, the utilities suffered as a result of inflation, high interest rates, heavy investment in new generating facilities (which often failed to perform at their rated capacity and were less reliable than expected), and traditional regulatory practices and procedures that produced the phenomenon of "regulatory lag." These factors reduced the profits of many utilities below their allowed rate of return. *See* S. Mintz, "An Explanation of Electric Utility Finance and Its Effects on the Residential Consumer" (Paper presented at the 22nd Annual American Council on Consumer Interests Conference, Atlanta, Georgia, April 8, 1976).

5. According to a Library of Congress estimate, the colder than normal winter in 1976—1977 would raise the average household heating bill by $139. Hobart Rowan and William Nye Curry, "Fuel Cost Increase of $139 Per Home Seen This Winter," *Washington Post*, February 3, 1977, p. A—1.

6. The electricity and natural gas rate data in this paragraph are from *Electric and Gas Utility Rate and Fuel Adjustment Clause Increases, 1977*, prepared for the U.S. Senate Committee on Governmental Affairs by the Congressional Research Service, Library of Congress (Washington, D.C.: U.S. Government Printing Office, 1978), p. vii.

7. The Natural Gas Policy Act of 1978, P. L. 95—621.

8. The price of new gas would increase from $2.06 per thousand cubic feet (mcf) on October 1, 1978, to $3.85 per mcf on January 1, 1985, according to

estimates of the U.S. House of Representatives Committee on Interstate and Foreign Commerce, Energy and Power Subcommittee. Environmental Study Conference (U.S. Congress), "Fact Sheet on the National Energy Act" (November 1978).

9. M.C. Barth et al., "The Impact of Rising Residential Energy Prices on the Low Income Population: An Analysis of the Home-Heating Problem and Short-Run Policy Alternatives," Technical Analysis Paper No. 3, Office of Income Security Policy (Washington, D.C.: U.S. Department of Health, Education and Welfare, March 1975), p. I–6. According to this study, the poor pay a higher percentage of their income, despite consuming only 56 percent as much electricity and 82 percent as much natural gas as the nonpoor.

According to the Consumer Federation of America, the poorest 10 percent of the population spends a ten times greater proportion of its income on residential energy than does the richest 10 percent. K.F. O'Reilly, " 'Immoral' Gas Compromise," *Washington Post*, August 17, 1978, p. A–23.

10. Mintz, *supra* note 4 at 24, citing Federal Power Commission, *Statistics of Privately Owned Electric Utilities in the United States* (Washington, D.C.: Government Printing Office, 1973), p. lxix. By September 1978, average rates for residential electricity consumers had risen to 4.48 cents per kwh; for commercial customers, to 4.41 cents per kwh; and for industrial customers, to 2.79 cents per kwh. Department of Energy statistics, reported in *Energy User News*, January 22, 1979, p. 16. The disparity in rates charged residential and industrial electric customers, despite having narrowed slightly, remains substantial.

11. Subsistence energy needs may vary widely from one household to the next. A subsistence level of electricity for a family in an all-electric home will obviously be much greater than the subsistence level for a family that also uses natural gas or heating oil to meet its needs. The difficulty in determining how much electricity is required for subsistence needs is reflected in the range in the amount of the lifeline block (300–700 kwh) in various lifeline proposals.

12. See notes 19–23, *infra*, for examples of these two approaches.

In designing lifeline rate structures, legislators and public utility commission officials should be aware that fuel adjustment clauses and customer service charges could substantially reduce lifeline benefits. To avoid this result, legislators and utility commissions should consider exempting the lifeline block from the fuel adjustment clause and allowing service charges to be imposed only on customers who use more than the lifeline amount.

13. *See* J.D. Pace, "Lifeline Rates and Energy Stamps" (Paper presented at National Economic Research Associates, Inc., Conference on Peak-Load Pricing and Lifeline Rates, New York, New York, June 17, 1975). ("Surely it is not the utility's responsibility to see that all groups in our society are provided with adequate incomes or sufficient price subsidies to enable them to obtain the necessities of life." *Id.* at 2.)

14. *See*, e.g., H.R. 11449 and 12461; and S. 3011, 3310, and 3311.

15. The Public Utility Regulatory Policy Act of 1978, P. L. 95–617, § 114 (b), 16 U.S.C. § 2624(b). This provision requires utility commissions to hold evidentiary hearings to determine whether electric utilities under their jurisdic-

tion *should* have lifeline rates if, by October 1980, the utilities do not have lifeline rates.

16. The sources of these figures are Mintz, *supra* note 4 at 25; and National Conference of State Legislatures, "Energy Report to the States," no. 76–11, Denver, Colorado (April 30, 1976).

17. For instance, following the enactment of the Miller-Warren Energy Lifeline Act (see note 20, *infra*), the California Public Utilities Commission, in Case No. 9988 (filed October 7, 1975), began an investigation on its own motion into a determination of lifeline amounts and rates for electricity and natural gas. The commission's decision (No. 86087) establishing the amounts and rates was filed on July 13, 1976.

An example of a commission considering lifeline rates as a result of citizen participation occurred in Montana, where in 1976 the Public Service Commission considered a lifeline proposal submitted by the Center for the Public Interest in Bozeman, Montana.

18. In Arkansas, the Arkansas Community Organizations for Reform Now (ACORN) successfully placed lifeline initiatives on the 1976 ballot in the cities of Pine Bluff (where it was defeated) and Little Rock (where it passed). After this electoral victory, however, the law empowering cities and towns to set utility rates was repealed, and a state trial court struck down the Little Rock lifeline ordinance on grounds unrelated to the validity of the lifeline concept (i.e., taking of property without just compensation and illegal delegation of authority). *See* Arkansas Foundry Co. v. City of Little Rock, No. 76–4893 (Ch. of Pulaski County, Ark. Feb. 4, 1977); and First Electric Cooperative Corp. v. City of Little Rock, No. 76–4959 (Ch. of Pulaski County, Ark. Feb. 4, 1977). These decisions were not appealed.

In Massachusetts, voters defeated a "fair share" rates (all users would have been charged the same rate) initiative in 1976.

In Ohio, Ohioans for Utility Reform sponsored a constitutional amendment in 1976 to provide lifeline rates for natural gas and electricity. Beseiged by publicity against the initiative from utilities and industrial users, voters rejected this proposed amendment.

In 1978, South Dakota voters rejected a lifeline proposal.

19. Ch. 585, S. L. 1975. Persons sixty-two years of age or older were eligible for lifeline rates of not more than 3 cents per kwh for the first 500 kwh used each month if the household income was less than $4,500 for a single member household or $5,000 for a household with two or more members.

20. Ch. 1010, S. L. 1975, codified at Cal. Pub. Util. Code 739. "The lifeline rate shall be not greater than the rates in effect on January 1, 1976. The commission shall authorize no increase in the lifeline rate until the average system rate in cents per kilowatt-hour or cents per therm has increased 25% or more over the January 1, 1976, level." § 739(b).

21. Ch. 440, P.L. 1977. "The lifeline rate shall be equal to or lower than the lowest effective rate [average per unit cost charged to any class of customers] per kilowatt-hour or per therm at which gas is sold to any class of customers of the utility. . . . " To be eligible for lifeline benefits, a "user" must have income

of less than $12,000 for a married couple or head of household or $9,000 for a single person. The lifeline burden (i.e., higher rates) will be borne by all classes of customers. The most interesting provision of the legislation authorizes the appropriation of revenues derived from the licensing and taxation of gambling casinos to establish and maintain a lifeline rate for eligible users who are sixty-five years of age or older or who are disabled.

22. Aztec, New Mexico's municipally owned electric utility has a lifeline rate for senior citizens on fixed incomes.

23. In the District of Columbia, Georgia, and Rhode Island, public service commissions have frozen rates for the initial block of kwh. In early 1978 the Minnesota public service commission ordered utilities to submit lifeline proposals and promised to implement one.

In addition to lifeline rates, which affect only residential customers, some public service commissions have ordered inverted rate structures (increasing rates for successive blocks of usage) for all customer classes. Inverted rates exist in Washington State (Seattle City Light and Puget Sound Power and Light), Michigan (Consumers Power and Detroit Edison), and Florida (Florida Power and Light).

24. *See* "The Lifeline Rate Concept," Office of Consumer Affairs/Special Impact, Federal Energy Administration (now U.S. Department of Energy) (undated); J.D. Pace, "The Poor, the Elderly and the Rising Cost of Energy" (Paper presented at the Pennsylvania Power Conference, Hershey, Pennsylvania, April 23, 1975).

25. For summaries of various studies on the relationship of electricity usage and income, *see* J.D. Pace, "Studies Examining the Relationship Between Electricity Usage and Income, Age and Other Factors" (New York, N.Y.: National Economic Research Associates, Inc., 1975), and "The Lifeline Rate Concept," *supra* note 24.

26. Pace, *supra* note 13 at 5.

27. *Id.*

28. *Id.* at 6.

29. Pace, *supra* note 24 at 10, n. 4.

30. Pace, *supra* note 13 at 4.

31. *Cf.* Cal. Pub. Util. Code § 739.5(a), which requires master meter customers who provide domestic gas or electric service to users through a submeter service system to "charge each user at the same rate which would be applicable if the user were receiving such gas or electricity or both, directly from the serving utility." This provision, in effect, insures that submeter users receive any lifeline benefits.

32. In April 1976 the New York Public Service Commission announced that it would prohibit the inclusion of electric service bills in rents paid by residential and commercial tenants in any building constructed after mid-1976. In addition to this action, the commission announced a review of its prohibition of submetering—the practice of building owners buying electricity at bulk rates and billing their tenants for use as measured on individual meters. Release 76130/F C. 26998 (April 25, 1976).

In 1977, Maryland, Minnesota, North Carolina, and Oregon passed legislation banning master metering of new buildings.

See also the Public Utility Regulatory Policy Act of 1978, P. L. 95−617 § 113(b)(1), 16 U.S.C. § 2623(b)(1).

33. "The Lifeline Rate Concept," *supra* note 24 at 24.

34. Pace, *supra* note 24 at 5−6.

35. *See* "The Lifeline Rate Concept," *supra* note 24 at 23.

36. Pace, *supra* note 13 at 6−7.

37. New York State Public Service Commission, Opinion 78−20 (August 30, 1978).

38. The commission stated: "We conclude that initial block discounts would be an ineffective means of dealing with the problem of the effect of rising electricity rates on the poor. They would impose rate increases . . . on many customers, including many low income families. And they would distort price signals, encouraging uneconomic use of electricity. The record here supports neither the belief that electric rates place a disproportionate burden on the poor, nor the hope that initial block discounts would reduce that burden. . . . [The record] shows that initial block discounts would injure some poor and benefit some affluent electric consumers." Opinion 78−20, Slip. op. at 19−20.

39. Energy stamp bills introduced in the 94th Congress included H.R. 16140 (also 16449 and 17076) and 17316. One bill introduced early in the 95th Congress was S. 609.

40. Project HELP Fact Sheet and Program Regulations, Community Action Committee of the Lehigh Valley, Inc., Bethlehem, Pennsylvania (1975).

41. "Stamps to Pay Fuel Bills," The Ann Arbor *News*, August 13, 1976, p. 11.

42. H. 142 (1975).

43. S.B. 849 (1975). In 1977 several bills were introduced in the Michigan legislature to create a utility services stamp program for senior citizens (H.B. 4046), senior citizens and public assistance recipients (H.B. 4371), and Social Security recipients (S.B. 559).

44. Memorandum, "Analysis of Senate Bill 849," from Michigan Department of Commerce to Governor William G. Milliken (March 17, 1976).

45. W.G. Rosenberg, "Impact of Higher Home Energy Costs on Low Income Families and Alternative Public Programs to Reduce Hardships" (Michigan Public Service Commission, June 25, 1975); and D. Johns and P. Proudfoot, "Public Subsidization of Energy Costs: Technical Analysis and Methodology" (Michigan Public Service Commission, June 30, 1975).

46. Florida Power Corporation, "Florida Energy Stamp Program" (St. Petersburg, February 1975).

47. Pace, *supra* note 24 at 13.

48. Rosenberg, *supra* note 45 at 19.

49. In addition to Connecticut, Oregon, and Kentucky, Wisconsin has an emergency fuel and utilities assistance program for low income households. Wisc. Stat. § 49.055.

50. Committee Bill 1004 (1975).

51. Public Act 75-3.

52. Public Acts 75-561 and 76-287.

53. The Emergency Assistance Program (42 U.S.C. § 606(e)) authorizes cash payments, payments in kind, and services to needy families, not necessarily AFDC recipients, with dependent children. To be eligible for this category of emergency assistance, the child must be "without available resources, (and) the payment, care, or services involved are necessary to avoid destitution of such child or to provide living arrangements in a home for such child." The assistance, which can be furnished only thirty days in any twelve month period, can be used to pay for any costs incurred during the twelve month period. Therefore, emergency assistance payments could be applied against future as well as past energy costs.

54. Connecticut State Welfare Department, Social Service Policies Manual, vol. 1, paras. 5050, 5060 (1976).

55. Public Act 78-184.

56. H.B. 3007 (1977).

57. Ch. 185, S. L. 1978.

58. The law requires the department of human resources to report to the interim joint committee for public utilities and transportation on July 1, 1979, and July 1, 1980, on the implementation of the program.

Appendix to Chapter 5

STATE OF MICHIGAN SENATE BILL 849
(as amended)

Senate Bill 849 was introduced May 13, 1975 and died in committee in 1976. The legislation in this appendix is a revision of Senate Bill 849 incorporating the changes suggested by the Michigan Department of Commerce to extend the coverage of the bill from senior citizens to all low income households. *See* Memorandum, "Analysis of Senate Bill 849" from the Michigan Department of Commerce to William G. Milliken, Governor (March 17, 1976).

A bill to establish an energy voucher program for low income households; to provide for the administration of the program; to prescribe powers and duties of the department of social services and the department of treasury; and to prescribe penalties.

THE PEOPLE OF THE STATE OF MICHIGAN ENACT:

Sec. 1. As used in this act:

(a) "Voucher" means a utility voucher or certificate issued pursuant to this act.

(b) "Coupon allotment" means the total value of coupons issued to a low income household during each calendar quarter or year.

(c) "Household" means any single individual occupying a dwelling unit or any group of individuals occupying a dwelling unit either related or unrelated.

(d) "Household income" means income as defined in section 510 of Act No. 281 of the Public Acts of 1967, as amended, being section 206.510 of the Michigan Compiled Laws.

(e) "Municipal utility" means a gas or electric utility company operated pursuant to chapter 12 of Act No. 3 of the Public Acts of 1895, as amended, being sections 72.1 to 72.9 of the Michigan Compiled Laws; chapter 27 of Act No. 215 of the Public Acts of 1895, as amended, being sections 107.1 to 107.10 of the Michigan Compiled Laws; Act No. 186 of the Public Acts of 1891, as amended, being sections 123.91 to 123.93 of the Michigan Compiled Laws; or, section 1 of Act No. 41 of the Public Acts of 1895, being section 123.101 of the Michigan Compiled Laws.

(f) "Public utility" means a gas or electric utility company regulated by the public service commission pursuant to Act No. 3 of the Public Acts of 1939, as amended, being sections 460.1 to 460.8 of the Michigan Compiled Laws.

(g) "Home space heating energy dealer" means a retail distributor of home heating oil or liquid propane gas.

Sec. 2. (1) The department of social services shall formulate and administer an energy voucher program under which eligible households may purchase energy vouchers to assist the households in paying the increased cost of home energy bills through the issuance of vouchers which shall have a greater monetary value than the price paid for the voucher by the household.

(2) The increased costs of home energy bills shall be computed using a minimum energy budget composed of 271 kilowatt hours per month and 167 thousand cubic feet of natural gas per year or the energy equivalent for home heating oil or liquid propane gas with the base price being the costs of these commodities in September, 1973.

(3) The vouchers shall be used to defray the increased costs of home energy and may be applied to one or all of the bills in whole or in part from a municipal or public utility or a home space heating energy dealer.

Sec. 3. (1) The department of social services shall issue an energy voucher only to households who are certified as eligible to participate in the energy voucher program.

(2) An energy voucher shall not be issued to:

(a) A household with an income in excess of $6,000 per year.

(b) More than one individual in each household.

(c) A person who resides in an institution.

(d) Any person living in a household which is not certified to participate in the program.

Sec. 4. (1) A household may make application to participate in the energy voucher program on forms prescribed by the department of social services.

(2) Upon verification by the department of social services that the household is eligible to participate in the energy voucher program, the department shall issue a voucher for that household. The certification shall be valid for not more than one year. The voucher shall specify the amount or benefit to which the household is entitled and the purchase price to be paid for the voucher.

Sec. 5. (1) A person shall not knowingly make a material false statement in an application to participate in the energy voucher program.

(2) The department of social services shall not disclose the information obtained from an applicant except as is necessary to carry out the energy voucher program.

Sec. 6. The department of social services shall promulgate rules pursuant to Act No. 306 of the Public Acts of 1969, as amended, being sections 24.201 to 24.315 of the Michigan Compiled Laws, to administer the energy voucher program.

Sec. 7. (1) The department of treasury shall issue energy vouchers.

(2) The vouchers shall be simple in design and shall only include the words or illustrations necessary to explain their purpose and define their amount.

(3) The department of treasury shall designate offices at which vouchers may be purchased. The offices shall be conveniently located through the state.

(4) Upon presentation of a valid certificate of eligibility at a designated office, the designated member of a household may purchase the voucher as authorized in the certificate.

Sec. 8. (1) A municipal or public utility or a home space heating energy dealer shall accept at full face value vouchers issued and purchased as provided in this act as payment for gas or electric or other designated home space heating energy rendered to a certified household.

(2) A municipal or public utility or a home space heating energy dealer may redeem at full face value the vouchers issued and used as provided in this act by depositing the vouchers with the department of treasury.

Sec. 9. (1) A household shall pay a charge for the voucher. The amount of the charge shall represent a reasonable investment on the part of the household but shall not exceed 5% of the annual household income.

(2) The revenues received from the charge made for the vouchers shall be deposited in the state treasury and credited to a special account. The revenues shall be reserved for the redemption of coupons pursuant to section 8 (2).

Sec. 10. The value of the voucher issued to a household which is in excess of the amount charged for the voucher shall not be considered income for any purpose under Act No. 281 of the Public Acts of 1967, as amended, being sections 206.1 to 206.532 of the Michigan Compiled Laws.

Sec. 11. A person shall not knowingly alter, acquire, use, or transfer a voucher in any manner not authorized by this act.

Sec. 12. A person who violates this act is guilty of a felony and shall be imprisoned not more than 5 years or fined not more than $20,000.00, or both.

Conclusion

In the first three months of 1979, a series of events reminded the nation of the energy problem it had been trying to forget since the end of the Arab oil embargo in 1974. On January 1, the Organization of Petroleum Exporting Counties (OPEC) imposed the first of several price increases designed to raise oil prices 14.5 percent by the end of the year. As political upheaval intensified in Iran, the world began to feel the effects of the loss of Iranian oil exports. To fill this void, other oil-exporting countries temporarily expanded their production, and prices on the "spot market" (the market for oil not sold under contract) soared. With demand high, several OPEC nations unilaterally boosted prices above the official level. After Iran resumed limited production and export in March, it found buyers willing to pay prices one-third higher than the OPEC price. When the OPEC nations met in Geneva in late March, they raised prices 9 percent effective April 1 and voted to allow individual members of the cartel to add surcharges to the new base price.

In the United States, the interruption in the flow of Iranian imports, which constituted 5 percent of total oil imports and 2.5 percent of total oil consumption, was felt when the oil companies began to allocate gasoline supplies to dealers. Government officials warned motorists that summer would bring gasoline shortages, that gasoline would cost at least 10 to 15 cents more by the end of the year, and that the days of $1 per gallon gasoline were in the immediate rather than the distant future. The Treasury Department estimated that the nation's bill for oil imports in 1979 would exceed $50 billion, up over 20 percent from 1978.

Amid conflicting pronouncements on the seriousness of the supply shortage and price increases, President Carter and federal energy officials urged the American people to adopt voluntary conservation measures, such as obeying the 55 mph speed limit and turning down furnace thermostats. With other members of the International Energy Agency, the United States agreed to reduce its oil consumption by 5 percent to ease the shortage and discourage further price increases. Recognizing that voluntary conservation efforts could not fulfill this commitment, the administration proposed saving oil by increasing natural gas consumption, "wheeling" electricity from utilities that rely on coal or nuclear power to those that depend on oil, and mandating building thermostat settings (80° in summer and 65° in winter). Spurred by the interruption in Iranian imports, the Department of Energy finally prepared three energy conservation contingency plans and standby gasoline rationing regulations and submitted them for congressional approval nearly three years late.[1] Required by the Energy Policy and Conservation Act of 1975[2], the rationing regulations and contingency plans, which would impose emergency restrictions on weekend gasoline sales, building temperatures, and advertising lighting, would be implemented if the United States experienced a "severe energy supply interruption." Finally, before the end of the year, the administration must make a number of tough energy decisions including how much and how quickly to allow domestic oil prices to rise when EPCA's price controls expire in June. Unlike requests for voluntary conservation, these decisions will be difficult and will involve weighing conflicting energy, economic, and political considerations.

Even as the federal government begins to demonstrate a tentative commitment to tackling the nation's energy problem, several developments suggest that its efforts may meet some resistance. From Detroit, the automakers have mounted a campaign to relax the fuel economy standards mandated by EPCA. Despite the benefits of the standards (i.e., cutting the country's dependence on expensive oil imports and the motorists' operating costs), the manufacturers argue that the schedule for the standards should be stretched out to reduce the costs of compliance. A much more disturbing development, however, is the campaign that has sprouted in several western state legislatures to repeal the national 55 mph speed limit. At a time when the country faces gasoline shortages, efforts to scuttle a gasoline-saving—not to mention life-saving—measure appear irrational and irresponsible.

Opposition to conservation measures such as the 55 mph speed limit, however irrational, reflects the still prevalent opinion that the nation's energy problem is not serious, not permanent, and probably the result of a conspiracy among selfish oil companies. Accepting these fantasies is more comfortable and less threatening than confronting the hard realities that the energy problem will not disappear and that it will alter the way we live.[3]

Regardless of what actions, such as decontrolling domestic oil prices, the federal government takes in the next several years to overcome misconceptions and deal with the energy problem, state and local governments can complement federal programs by adopting policies to promote energy conservation. The economic disincentives analyzed in this book are one set of tools through which states and localities can encourage people to save energy. By serving as laboratories for the testing of economic disincentives, states and localities will earn themselves a place in the vanguard of the effort to solve the nation's energy problem.

NOTES TO CONCLUSION

1. The Energy Policy and Conservation Act, P. L. 94–163, §§ 201–203, 42 U.S.C. §§ 6261–6263, required the president to prepare and submit to Congress energy conservation contingency plans and a gasoline-rationing contingency plan by June 1976. In March 1979, the Department of Energy issued three contingency plans, covering weekend gasoline sales, building temperatures, and advertising lighting, 44 Fed. Reg. 12906–12917 (March 8, 1979), and standby gasoline rationing regulations, proposed 10 C.F.R. Part 570, 44 Fed. Reg. 15568–15597 (March 14, 1979).

2. The Energy Policy and Conservation Act, P. L. 94–163, §§ 201–203, 42 U.S.C. §§ 6261–6263.

3. *See, e.g.,* U.S. Congress, Office of Technology Assessment, *Changes in the Future Use and Characteristics of the Automobile Transportation System* (Washington, D.C.: U.S. Government Printing Office, 1979), which warns that as demand for oil surpasses production capacity in the 1980s, the nation's heavy reliance on the automobile will be seriously threatened. *See also* Executive Office of the President, Council on Environmental Quality, *The Good News About Energy* (Washington, D.C.: U.S. Government Printing Office, 1979), which reports that by increasing energy productivity, the nation can have a healthy, expanding economy with smaller than expected growth in energy consumption (i.e., a 25 percent versus 100 percent increase over current energy consumption by the year 2000).

Index

About the Author

Joe W. Russell, Jr., staff attorney at the Environmental Law Institute, is a graduate of Duke University and Georgetown University Law Center. In addition to his work on the Energy Conservation Project, Mr. Russell has studied alternative approaches for compensating victims of toxic substances pollution and analyzed the implementation of energy-efficient procurement practices in state and local government. He is currently helping prepare a handbook for homeowners and their lawyers who are challenging restrictive covenants that bar the installation of solar collectors.

DATE DUE

GAYLORD

PRINTED IN U.S.A.